The ARRL Antenna Compendium
Volume 1

Editors
Gerald L. Hall, K1TD
Paul Rinaldo, W4RI
Maureen Thompson, KA1DYZ

Contributors
Those 31 authors whose material
is published herein

Production
Maureen Thompson, KA1DYZ

Cover Design
Sue Fagan

Published by the
American Radio Relay League
Newington, CT USA 06111

THE COVER: A fall sunrise silhouettes the antennas of station W1AW.
(Photo by Jerry Hall, K1TD)

Foreword

The topic of antennas is one of the most popular in Amateur Radio literature. The material that has been written for amateurs on antennas would fill several volumes, and, indeed, several volumes are already available. So popular is the topic that ARRL Headquarters receives many more antenna manuscripts for consideration as feature article material for the journal, **QST**, than can be published. In the past we have had to turn away good material because of space limitations in the journal.

Instead of returning that material to the authors unpublished, why not collect it and publish it in a single volume? With this thought, **The ARRL Antenna Compendium** was born. Additional material was solicited for Volume 1, and you have the result in your hands.

Early in the planning stages, we chose to use the format of **QEX** in preparing material for publication. The authors have provided their own camera-ready art work, and "typesetting" has been done with Apple IIe computers and an NEC-3550 letter-quality printer. This approach was an experiment; we hope you will like the results.

There is a wealth of material between these covers, and on a variety of antenna subjects. If you have a serious interest in antenna design or construction, you'll likely find something here that is "right up your alley." At this writing we anticipate that the **Compendium** will become quite popular, and that Volume 2 will be a logical follow-on. We'd appreciate receiving your comments and suggestions for the next volume.

David Sumner, K1ZZ
Executive Vice President
Newington, CT

Contents

Vertical Antennas

Antennas of Reduced Size

Miscellaneous Antennas

Antenna Construction and Installation

General Antenna and Transmission-Line Information

Quad and Loop Antennas

The Sloping Diamond, A Full-Wave Loop for Four Bands the Easy Way

Duane R. Sanderson,* WØTID

*3735 S. E. Stanley Rd., Tecumseh, KS 66542

The full-wave loop antenna has become the subject of increasing interest among antenna experimenters in recent years. Judging by the volume of questions and comments I receive on the air, the loop is gaining popularity as an above-average performer. After 34 years as an active licensed amateur in pursuit of "that one great antenna," I think I have tried just about all applications of the standard quarter and half-wave antenna.

My experience has shown that folded-dipole antennas have always performed better for me than standard dipoles. The folded dipole is basically a folded full-wave loop. An argument as to which antenna is best is not the point here. I only mention this to identify the experimenting trail that led me to an antenna I currently use. I decided my search for a better antenna would be in the full-wave loop category.

Research Before Construction

A review of published antenna articles, daydreaming and mental pictures was necessary before I began to string wire. Published data on loops stated that some of their advantages are to provide gain over a dipole, and they are usually quiet even in a noisy environment. More broadbanded than half-wave types, they will work on more than one band.

These good points are a match for my antenna requirements. My QTH is in a rural setting with leaky high tension power lines at the front of the property. I work both SSB and CW, and wanted an antenna that would work on 40 meters with 15-meter capability, and be coaxial line fed. I decided to put up a full-wave loop. Now, what's next?

Loop Shape

Different handbooks state that loops can be circular, square, rectangular, delta or diamond. My use of this term "diamond" differs from the square in that the two inside distances of a diamond, between the sets of opposite points, are not equal. It is longer than it is wide.

From a construction standpoint, I concluded the delta was the most practical shape. Only one mast is needed to support the apex of a vertical delta, and metal fence posts can be driven into the yard at appropriate places to tie off the two bottom points.

It just so happened that I had a 45-foot mast near the garage with a TV antenna on top. The center of the 40-meter dipole was attached about one foot below the TV antenna. The dipole was fed with a balun and a 52-ohm mini-foam coaxial cable. I lowered the ends of the dipole and connected the appropriate half-wavelength of wire to them to form

the loop. The formula I used for the loop length is the standard L = 1005/f (MHz). I used 142 feet of no. 12 gauge copper wire for my loop that I stretched into a vertical delta shape, tying the bottom corners to appropriately spaced 4-foot metal fence posts. I could just reach the lower horizontal part of the delta while standing on the ground. The appropriate prune-and-tune period took many trips from the rig to the antenna and back again with the best SWR figure of 1.6:1.

Performance was a bit disappointing. Orientation was broadside east and west. Forty-meter daytime short skip was not as good as on the old dipole, and signal reports often included heavy "QSB" and "fading" comments. Long skip and nighttime

Fig. 1 — The sloping-diamond loop antenna.

performance was somewhat improved, but not significantly better than the dipole. The delta did not perform well on 20, 15, and 10 meters.

After several weeks of delta operation, I elected to try another loop shape. Again, the new shape had to work using only one mast for primary support, so I pulled the delta into the shape of a diamond and tied the two side points and the lower point to nearby objects. After pruning and tuning, I found the SWR was 1.3:1. I was making progress in obtaining a match, at least.

Vertical vs Horizontal

At this point my quad loop was a vertical diamond, broadside to the southeast and northwest. It was fed at the top, using a 1:1 balun and a 52-ohm RG-58 mini-foam coaxial cable. The total length of the loop wire was approximately 142 feet.

Daytime reports were fairly good but not spectacular. However, nighttime reports were significantly better than any wire antenna I have had at this QTH. The diamond outperformed my half slopers, which are mounted on a 50-foot tower, by at least one S unit or more.

The next few days and evenings were spent on 40-meter CW enjoying the short-skip daytime 579 reports, and the nighttime 589 to 599 reports from the West Coast. The power input was 60 watts. Operation of the loop on 20 CW resulted in rather poor performance compared with a 20-meter dipole on the tower. I concluded my diamond loop to be a good 40-meter nighttime long skip and DX antenna.

As time progressed, I encountered a few stations in nearby states who were using horizontal quad loops. These stations were consistently loud on the band in the daytime, and seemed to be more fade-free than others. They seemed to be using an antenna with high-angle radiation with reinforcement from nearby ground reflection, and had the equivalent of a two-element quad pointing straight up.

In choosing this antenna for use at my QTH, I recognized that most of my operating is in the daytime or early evening hours, with late-night DXing an occasional event. This is probably true for the majority of amateurs on 40.

I found that, still using the 45-foot TV mast as the top of the diamond, I could pull the loop away from the vertical plane to nearly horizontal, but with some slope. I placed three metal fence posts in appropriate locations in the yard as tie points for the diamond shape. My horizontal quad became a loop sloper with an approximate 30-degree slope off horizontal. See Fig. 1.

Once I used the prune-and-tune method on my new antenna, these were the results I obtained. At 7.1 MHz no reflected power was measured (1:1 SWR). The horizontal-loop impedance was a perfect match to the RG-58 coaxial cable and the broadband characteristics are the best I have ever seen in any of my antenna projects. A check of the SWR on each band from 40 through 10 revealed the following:

7.0 MHz	7.1 MHz	7.2 MHz	7.3 MHz
1.2:1	1:1	1.2:1	1.5:1

14.0 MHz	14.1 MHz	14.2 MHz	14.3 MHz
1.3:1	1.1:1	1.2:1	1.6:1

21.0 MHz	21.1 MHz	21.2 MHz	21.3 MHz
1.8:1	1.7:1	1.9:1	2.0:1

28.0 MHz	28.1 MHz	28.2 MHz	28.3 MHz
1.9:1	1.9:1	2.3:1	2.2:1

Does It Work?

You bet it does, and it works well on more than one band. Daytime 40-meter operation is great. Reports received while using 60 watts input are usually 599, and often there are reinforcing comments or questions from the other stations about my antenna. The receiving performance of the sloping diamond is equally good and demonstrates the quiet nature of loops, with low noise and loud signals. Nighttime operation is similar to daytime operation until late evening when West Coast stations overtake the band. At that hour, my half slopers are about two S units better than the loop. This is because of their low-angle characteristics. The bonus came when the loop was used on 20 meters. The SWR was almost as good as 40 meters, and performance on 20 is generally better than my dipole on the tower. Performance on 15 is almost as good as the 3/4-wave slopers on the tower. Performance on 10 has not been evaluated as of this writing because the band is generally dead, a victim of the sunspot cycle.

My approach to operating with this antenna is to use a line tuner on 21 and 28 MHz to tune out the small amount of reactance present on those bands, and to provide an ideal match to the rig.

The Clincher

The sloping diamond was put through a good test a few weeks after it was erected. The world-wide QRP contest arrived on the calendar and about mid-morning of the first day, I decided to hook my Argonaut to the loop and have a go at it. I spent about 2-1/2 hours working the 40-, 20-, and 15-meter CW bands. I worked 50 stations without difficulty. Among my contacts were G4, KH6, KL7, VE, and a good spread of stations across the U. S. A.

This antenna appears to be omnidirectional. At present, I am unable to find holes or blind spots in its performance.

Some Construction Comments

A rectangular horizontal loop has four corners to support, but the use of one mast as the primary support keeps things simple. A point on an existing tower, building or other structure can serve equally well. The mast I use is made of steel conduit, 1-3/4-inch diameter at the bottom. Successively smaller diameter sections are telescoped together a few inches at each joint with bolts and sheet metal screws through the overlapping points. I have found that sheet metal screws will help maintain a good electrical bond between each mast section. The base of the mast is hinged and set in a small poured concrete base for stability. My need for mast hinging stems from a lot of antenna experimentation at my location and with this setup, I can work alone to raise or lower the antenna. The mast hinge is made of two 16-inch pieces of 1-1/2-inch iron pipe set vertically with 10 inches in the concrete and 6 inches exposed. The two pipes are about 1-3/4 inches apart with sufficient space for the mast to set between them. All three pipes are then drilled so that a long bolt can pass through all three, as shown in Fig. 2. The bolt is the hinge or pivot point. My mast has one top guy on the side opposite the antenna, and the diamond shape of the antenna forms the equivalent of two other guys.

The end product is a 3-guy arrangement, and the stability of the mast is surprisingly good. A house bracket about midway up the mast ties it to the eaves of the roof. This makes the mast rigid and flexproof.

Summary

The unique propagation advantages of a horizontal loop seem to combine, to some degree, with the propagation advantages of a vertical loop when the plane of the loop is sloping. The amount of ground space required for the fence-post anchors is determined by the degree to which the constructor pulls the diamond sides apart. My diamond is nearly twice as long as it is wide. **The ARRL Antenna Book** states that next to a circular loop, a square loop is the most efficient, and rectangular loops or unequal diamonds have some energy cancellation. My

loop is more broadbanded and provides the best match to 52-ohm coax, when it is shaped as an unequal diamond. The use of a balun is optional. My use of balanced feed stems from my preference for a symmetrical pattern.

My experience with this antenna as a sloping loop has produced reports that are considerably better than when the loop is hung vertically. The exception is for late-night long skip where a vertical loop would perform best.

My home is surrounded by hilly terrain that is 40 feet plus higher than my yard elevation. Amateurs with low lying antenna sites, or at locations surrounded by tall objects, may find the horizontal or sloping loop to be the best choice. Those with large lots may want to try a really big loop and go for an 80-meter loop or one for 160 meters.

Soon I will have an 80-meter diamond sloper strung from a mast on top of one of those hills I mentioned. Listen for me on the bands and help me find out how well it works!

Fig. 2 -- A mast hinge is made of two lengths of 1-1/2-in iron pipe set in concrete. Their 1-3/4-in separation accommodates the steel conduit of the homemade telescoping mast.

Optimum Gain Boomless Quad

By Harold T. Mitchell,* NØARQ

*2403 Inca Lane, New Brighton, MN 55112

This 2-element beam can be precut for 10, 15, and 20 meters, with provision for 12 and 17 meters. It can also be adjusted for maximum front-to-back ratio using mini-stubs.

Perhaps you are interested in a quiet beam antenna, superior performance at low heights, single coaxial cable feed without an antenna tuner, high quad strength through cross ties, and a one-man installation of a 3-band quad. This article tells how those were obtained, plus optimum gain spacing for each band, and maximum front-to-back ratio adjustability. The antenna material cost a total of $75, and a guyed 13-foot roof-top mast was purchased for $45.

History and Background

The common 2-element 3-band concentric quad was patented by James C. McCaig of England in 1960.\1 It used bamboo canes and had an 8-foot boom for 10, 15, and 20 meters. Eighteen years earlier, Clarence C. Moore, W9LZX, had invented the cubical quad antenna to solve corona problems encountered at the 9,500 foot altitude site of HCJB in Quito, Ecuador.\2 A beautiful picture of four 6-element HCJB quads against the Andes was shown on the cover of a recent broadcasting magazine.\3 Quads have withstood the test of time.

The large thin planar elements of either a single or multiband quad have trouble resisting high bending loads. R. Michael Doherty invented a quad with spreader-reinforced crossarms, stating in his patent, "The chief disadvantage to cubical quad antennas is their lack of strength and vulnerability to high winds and icing conditions."\4

In New Brighton, MN, we are faced with severe weather, plus a restrictive antenna height ordinance. I turned to the quad for its superior performance when installed at low heights, and picked up the challenge of designing a strong quad. At my location, the 20-m loop bottom wire is only 1/4 wavelength above ground, yet the vertical angle of radiation for 20 meters is still below 40 degrees.\5

I designed this quad with not only performance and strength in mind, but also simplicity, one-man installation, and low cost. Parallel invention, simultaneous development, or redevelopment are not uncommon in history. When almost done with the detail design based on a local quad, I investigated the literature and found the Gem Quad antenna invented by Emerson G. Partridge, VE4RA.\6,7 The Partridge patent describes a boomless quad with unique fiberglass construction having inherently low wind resistance. My design uses Sitka spruce for spreaders. For the ham willing to construct a beam antenna, this design offers several new advantages, plus a rock bottom price well below that for even a bamboo quad (Fig. 1).

Why Boomless?

Quad wire loops can be square, circular or some intermediate shape, but they have one thing in common: All are planar. In the case of a 3-band concentric quad, the three loops of each element are also in the same plane. Wind and ice loads produce high bending stresses in the spreaders; high-strength fiberglass tubes, and even vaulting poles, are resorted to.\8

Boomless construction and 12 cross ties turn the quad into a structure with rigidity. Loads can be shared and mechanical oscillations eliminated. In addition, the spreaders are stabilized against side loads and high bending moments. The boomless quad is strong through structural design, and spruce spreaders can be used without undue concern. The quad contains 36 triangles, which are cross-coupled in three dimensions at each vertex.

Spiders were designed with 18-degree angles to obtain optimum gain 0.12-wave spacing for each band.\9 A concentric quad only can provide optimum gain for one band, since the boom length determines common spacing for all bands.

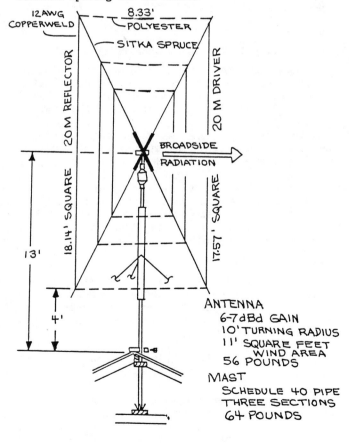

Fig. 1 — The end view of the roof-mounted 10-, 15-, and 20-m boomless quad. Each band has 0.12 wavelength optimum gain spacing.

Precut or Tunable?

Table 1 gives precut wire loops and spreader attachment points for five bands. The emerging 12- and 17-meter bands are covered in addition to 10, 15, and 20 meters.\10 There are a number of precut loop designs with booms for single band or concentric quads.\11,12 Precut antennas are convenient because construction and tuning steps are eliminated, which may be intimidating or burdensome. However, a considerable performance penalty is paid for this convenience.

R. J. Eckersley, G4FTJ, has stated an excellent case for antenna tuning, "For working DX it is clear that good front-to-back ratio is more important than forward gain — often the limiting factor in copying a weak signal is interference coming from the opposite direction..."\13 Bill Orr, W6SAI, adds weight to the need for quad tuning as he writes, "...the front-to-back ratio of the array is quite critical as to stub placement."\14 There are problems with most tuning stubs as they are relatively long and flimsy affairs, with 34- to 38-inch stub length for a 20-m reflector loop.\15

I have added 3 percent to the driver length to obtain precut reflector loops for the boomless quad. This is the accepted relationship that goes back to the work of Lee Bergren, WØAIW, in 1963.\16,17,18 The reflector loop lengths of Table 1 are also used for the tuned reflector quad. This permits small 12-inch x 3-inch mini-stubs rather than long stubs used where driver and reflector loops are initially of equal length. A tuned boomless quad is said to give an excellent 25-dB front-to-back ratio on 10 and 20 meters, and a very good 20 dB on 15 meters.\19

Relays, Baluns, Line Transformers, or Gammas?

A boomless quad requires solution of the basic antenna problem — how best to maximize radiated power and minimize SWR at the transmitter. Multiband quads have been matched to coaxial feed line by several different methods.\20

A tri-gamma matching system has been devised by Jack A. McCullough, W6CHE, and applied to a precut 4-element concentric quad with a 30-foot boom.\21 I have adapted the tri-gamma match (Fig. 2) to the optimum-gain boomless quad. A reactance capacitor balances gammas and the open-wire transmission line, which is the key to the system. The open-wire line connects all driver loops and the single coaxial cable from the transmitter.

Detail Design and Construction

The heart of the boomless quad consists of two short steel pipe spider hubs and a pipe axle mast T (Fig. 3). Welded to the hubs are steel angle stock spider arms to support the spreaders. With this hub and axle construction, quad elements can be installed one at a time (Fig. 4). The pair of elements then rotate easily in the vertical plane for complete cross tying (Fig. 5) prior to bolting to the hub axle.

If a diamond quad is preferred instead of a horizontal quad, it is only a matter of changing hole locations at the hub axle. For a diamond quad, the signal will be polarized at 45 degrees and contain both vertical and horizontal components unless the feed point is shifted to an apex. To obtain a vertically polarized signal, the quad is rotated to obtain feed on a vertical side.

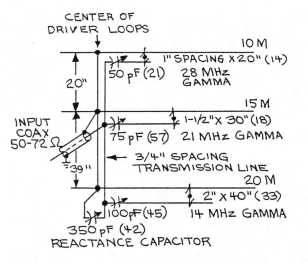

Fig. 2 -- The tri-gamma matching system provides impedance matching and reduces interaction between loops. Open-wire transmission line and gamma rods are 12 AWG solid. Back-to-back alligator clips are used for gamma rod length adjustment during matching. Values in parentheses are for RG-8 coaxial cable and NØARQ low elevation.

Precision assemblies follow from large simple welding fixtures glued together from 3/4-inch plywood. Spider accuracy results from a square hub face in contact with the fixture surface (Fig. 6A) and equally spaced 18 degree (5-3/16 inch x 16 inch) ramps to guide the steel angle. Axle T accuracy is obtained from a 3/4-inch slot, which lifts and aligns the mast pipe for welding to the larger hub axle pipe resting in a 1-1/2-inch slot at right angles (Fig. 6B).

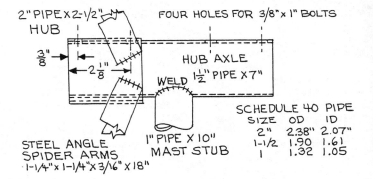

Fig. 3 -- The spider hub and axle mast T are the key to this boomless quad. Construction allows separate installation of each element as well as rotation together in the vertical plane for cross tieing.

Sitka spruce was selected for spreader material for its strength and low cost. Clear boards without knots come from these very tall trees that grow along the coast from Alaska to California. Sitka spruce is still used for propellers and framing small aircraft. Vertical grain boards are twice as expensive as flat grain and are unnecessary as the original grain direction is lost when ripped into a square cross section. Ten 7/8-inch x 7/8-inch spreaders were obtained from two 1-inch x 6-inch x 14-foot boards, with several thinner strips left over. Each spreader cost $1.76 and weighed 2.4

Fig. 4 — NØARQ slips the reflector element onto the mast T. A 63-inch spruce tie bar supports 12-inch x 3-inch mini-stubs in the bottom center of each reflector loop.

Fig. 5 -- NØART stands on the roof to add 12 polyester cross ties by rotating the quad in the vertical plane. Polyester is strong and stretches very little. It is also highly resistant to sunlight.

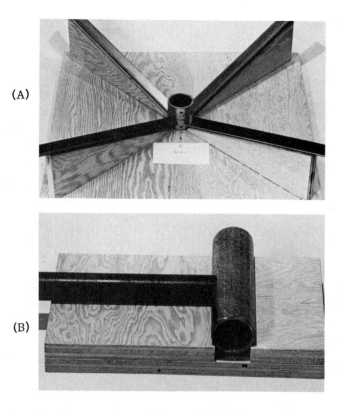

(A)

(B)

Fig. 6 -- At A, the spider welding fixture is constructed on a 2 foot x 2 foot x 3/4-inch plywood base. Evenly spaced 5 3/16-inch x 16-inch ramps provide the 18-degree spreader angles. At B, a hub axle and mast T fixture lifts the 1-8 inch mast pipe for welding at right angles to the 1-1/2-inch axle pipe.

pounds. A sharp saw blade left slightly rough surfaces, which were then primed and painted with light blue latex house paint to blend with the sky. Sanding or shaping of spreaders is not necessary.

Number 12 AWG solid, steel-core copperweld was used for strength and resistance to stretching. If all copper wire were chosen, stranded would be used for added resistance to breaking.\22 Spreaders were measured according to Table 1 and drilled on the diagonal for the 0.08-inch OD wire. It was nice working with solid wood of constant cross section rather than thin tapered bamboo or fiberglass cylinders. After threading and positioning the wires at each spreader, wrap wires were added and soldered to prevent shifting of the spreaders on wire loops.\23 Silicone rubber sealant was applied at drilled holes to prevent water entry and give electrical insulation.

Each spreader was designed to butt against its spider hub and not require adjustment. Spreaders were fastened to spider arms with glass filament shipping tape instead of the usual 16 stainless steel hose clamps (more costly). The shipping tape was then over-wrapped with black vinyl tape to protect it from sunlight.

It was necessary to plan ahead for placing various gamma-rod spacing insulators on driver loops before closure. To avoid breakage, the copperweld should be spliced and soldered without sharply bending wires.\24 Heavy scissors were used to cut out the open-wire line spacers from 1/16-inch Lexan polycarbonate sheet. Holes can be drilled or punched in this excellent electrical insulation, which is also tough and resiliant. The spacers were laced in place with nylon cord.\25 This is a very rapid, effective and economical method of building open-wire line (Fig. 7). It proved especially useful since five different spacings were required. As the first spacer of each reflector loop stub, an egg insulator was used to carry loop tension under load.\26

Fig. 7 — Coaxial cable input is at the 15-m point to the tri-gamma matching system. Open-wire line polycarbonate spacers were laced in place with nylon twine. Variable capacitors used during matching were removed, measured, and replaced with silver mica fixed capacitors. Coaxial braid and ends were coated with silicone rubber sealant to prevent moisture wicking, corrosion and change.

Final Assembly

The boomless quad was assembled on a home built, roof-top mast designed to withstand a 40 pound per square foot (100 mi/h) wind load on the system.\27 A mast height of 13 feet was chosen as a minimum to be consistent with the 12.8-foot vertical turning radius of the quad. The basic mast was fabricated from two 1-1/2-inch schedule 40 steel pipes. A 10-foot length of pipe weighing 27 pounds was the heaviest item lifted during installation of the quad and mast.

A 6-foot sleeve of 2-inch schedule 40 pipe was centered and fastened with 5/16-inch eyebolts in the middle of the mast to pin the two 1-1/2-inch pipe sections. The main function of the sleeve is to provide additional bending strength in that critical length. The mast, rotator, and tee were preassembled for fit and then finished with Rust-Oleum^{TM} primer and paint before taking sections to the roof for reassembly.

Each of three evenly spaced 3/16-inch EHS galvanized steel guy wires was brought down at a 45-degree angle from the eyebolts and terminated in a thimble.\28 A bridle through the thimble (Fig. 8) divided the load of each main cable between two 5/16-inch eyebolts screwed into the house frame. All cable ends were secured with 3/16-inch cable clamps.

Turnbuckles are unnecessary in this application, and can even be undesirable.\29

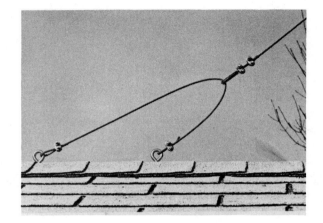

Fig. 8 — A bridle running through a thimble in each guy wire divides the load between two eye bolts screwed into the house frame. Two cable clamps are used at each end of the 3/16-inch EHS guy cable. One clamp is sufficient for each end of the 3/16-inch bridle cable.

From inside the roof, an entrance point for the 1.90-inch OD pipe was selected so that the pipe would rest against the roof center beam and a rafter (Fig. 9A). A center-line hole was drilled up through the roof and a plumb bob was dropped from a nail in the hole to locate the exact center of the mast base. A 2 x 8 was fastened to the joists above the ceiling and a circumferential pattern drilled for 8 nails, which form a funnel socket (Fig. 9B) for the mast. This may appear unorthodox, but it is very simple, and the shearing strength of three nails for any wind load direction is certainly sufficient. A corner-notched 2 x 8 was used to wedge the mast against the center beam and rafter. A roof-top mast clamp to prevent rotation was made from a V cut 2 x 4 (Fig. 10), and it was installed after sealing the roof entry.

There are other ways to roof mount an antenna, but this mast proved very strong, easy to build, and economical. It provided a precisely vertical mast without adjustment.\30,31 After the mast was assembled and guyed, each quad element was lifted into place. Then the elements were rotated together vertically and cross tied with 1/4-inch braided polyester cord at the 20-m loops and 3/16 inch for the 10- and 15-m loops. Polyester (not polypropylene) was used for its superior UV resistance and minimal stretch when compared with nylon. Although this antenna and mast were value engineered to reduce cost and overdesign, higher than normal quality materials were used where system strength and stability could be enhanced. After bolting spider hubs to the axle (Fig. 11), the antenna and rotator were aligned with true north, and final electrical and ground connections were made to the antenna, rotator and mast.

(A)

(B)

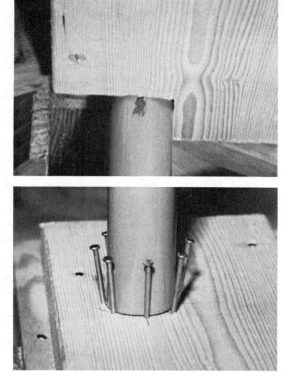

Fig. 9 — At A, the mast rests against the roof center beam and a rafter. A notched 2 x 8 completes a tight box below the roof. At B, a funnel pattern of nails in a 2 x 8 serves as mast socket and base.

Fig. 10 — A V-cut 2 x 4 and screw-adjustable block make a roof-mounted mast clamp to prevent rotation. The roof was sealed before clamp installation. A heavy ground wire is attached to the mast for lightning protection. Dressed along the mast are coaxial and rotator control cable, which has been wrapped with grounded copper mesh to prevent RF pickup from feeding the shack.

Tuning the Boomless Quad

1.   Set the gamma system broadcast type temporary variable capacitors and rod lengths approximately to the values found by NØARQ (Fig. 2). Since loops and adjustments interact, it is recommended that these initial values provide a starting point and that the following steps be taken

Fig. 11 — An end view of the quad shows pre-cut spreaders against the hub. The hub and axle pipe are self-aligning. The quad is symmetrical with low rotational inertia and presents modest loads to the rotator. Coaxial cable was brought up over the mast T for strain relief. A 7 foot 2 x 2 stiffener was fastened to each upper spreader with shipping tape overwrapped with black vinyl tape for protection from the sunlight.

carefully and sequentially. If tuning stubs are used, adjust each reflector loop stub for maximum front-to-back ratio. This can be done with a receiver S meter at the antenna, or with earphones on a long cord. Listen for a distant signal with the back (reflector) of the quad aimed broadside at the source. Move the alligator clip shorting bar until a minimum response is heard. If the 20-m reflector loop is too long with the stub length entirely shorted, then bridge the loop bottom corners an equal amount and fine tune with the stub. Because of local reflections, it may be desirable to try signals from more than one direction. The stubs are adjusted first, as these reflector adjustments will affect driver loop matching.\32

2.   Check that the shield of the input coaxial cable is connected at the 15-m level to the side of the open-wire transmission line that connects the center of each driver loop, and check that the reactance capacitor is at the 20-m level. Apply low power on 10 meters and adjust the 10-m gamma capacitor and rod length for lowest SWR as the transmitter frequency is varied, to find the band resonant frequency.

3.   Switch to 15 meters and tune the 15-m gamma for lowest SWR at resonance.

4.   Switch to 20 meters and tune similarly. Now also vary the reactance capacitor as well as the gamma capacitor and rod to further reduce the 20-m SWR.

5.   Recheck 15 meters, then 10 meters, and readjust for lowest SWR. If adjustments and results are not adding up, it will seem as if the SWR improves as gamma lengths are shortened. If the gamma lengths are much shorter than noted in Fig. 2, it is a good indication that the reactance capacitor setting is incorrect.

If band conditions are not cooperating to deliver distant signals when front-to-back adjustments are to be made, it may be desirable to enlist the services of a local ham. Your amateur friend should transmit a horizontally polarized signal from more than 500 feet to avoid confusing results.

This adjustment sequence produced excellent matching as measured by a Drake MN7 on the end of 35 feet of RG-8. SWR below 2.0 for two bands and the CW portion of 10 meters is possible because of the low Q characteristic of a cubical quad. From a loss standpoint, an SWR below 2.0 is considered fully acceptable.\33,34 It also permits full output of solid-state transmitters that automatically reduce output for SWR in excess of 2.0.

The four variable capacitors were removed and measured, then substituted, with 500-WVDC silver micas. Jumper wires were soldered in place of alligator-clip shorting bars. This quad is easy to work on, as all ties and adjustments are within arm's reach from either the roof or only the mast.

From the data (Fig. 12), very broadband performance is documented. With such low SWR, the boomless quad can be fed directly without an antenna tuner. An often overlooked benefit is also realized. The deleterious effect of antenna tuner Q on bandwidth is also eliminated.\35 If the transmitter frequency is changed without retuning an antenna tuner, then the tuner Q reduces power transfer to the antenna. Other things being equal, a directly matched antenna will have the greatest bandwidth.

Conclusion

The excellent performance of this antenna on the bands means that we have joined the big leagues. The entry fee was a modest dollar expenditure, plus time. For this investment, there has been the satisfaction of building a multiband quad that is the equal of 2-element commercial quads. I also believe that it is superior to the 3-element Yagi. It clearly has better bandwidth and front-to-back ratio, and is within tenths of a dB on forward gain. It wins on gain, too, if the Yagi is tuned for maximum front-to-back ratio.\36 At low heights there is no contest as the quad wins hands down for low-angle power radiation.

The 11 square feet of wind area causes little concern. Rotators are rated for unsupported antennas within 24 inches. The center of the boomless quad is within 6 inches. It is also a symmetrical structure with relatively low rotational inertia. Rotator drive and stopping requirements are modest.

With this roof mast and guying design, a total system with high strength was obtained. The 13-foot mast should handle practically any antenna and is recommended for a neat roof-top installation. Structural design and 12 AWG copperweld, together with quality materials and assembly, should solve the old strength problems of 2-element quads. The lace-like appearance of this antenna and mast will also draw lots of attention. One said it looks like art.

Appendix

A massive storm with 100 mi/h+ winds ripped through the Minneapolis — St. Paul area during the summer of 1983. It produced the most extensive local power failure in history. The mast stood but one spreader and several spider angles failed. The heavier spider steel angle stock of Fig. 3 gives a factor of 6 increase in bending strength. The top spreaders have been stiffened from the hub to above the 10-m wire, using 1-1/2 inch x 1-1/2 inch x 7 feet 2 x 2s. These were taped to the spreaders to give a factor of 5 increase in bending strength. As a result of the storm, an even stronger antenna was rebuilt and is reflected in this article.

With heavier steel and new top stiffeners, I now recommend against one person lifting elements into place. Moment arms magnify actual weight. Although possible, it becomes difficult for one person to align an element to slip its hub on the T. With a gin pole it would be an easy one-man job.

Acknowledgement

I wish to thank my son, NØART, for getting me started on beam antennas. Todd found a new HAM-M rotator (we have added our own brake variable delay modification) for sale at his Junior High School and pointed out that the quad is an unusually effective low-angle radiator when installed at low heights. Thanks also to Todd's brothers Tom, NØBSY, for improving this article, and Tore for taking and printing the photographs. I have appreciated the patience and tolerance of my wife during the period

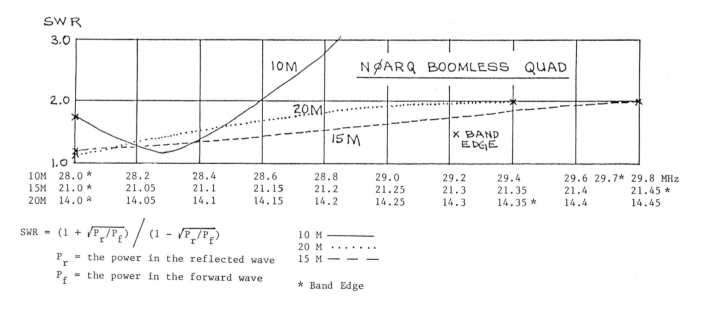

SWR = $\left(1 + \sqrt{P_r/P_f}\right) \Big/ \left(1 - \sqrt{P_r/P_f}\right)$

P_r = the power in the reflected wave

P_f = the power in the forward wave

Fig. 12 — SWR plots of NØARQ boomless quad.

of intense activity required to design, build, and publish something of worth.

R. P. Gruetzman, KØKLH, W. W. Karjalahti, WAØODW, and B. W. Schultz, WAØOU, are credited with developing the basic quad structure.

References

\1 McCaig, J. S., "Improvements Relating to Composite Aerials," U.K. Patent No. 850,974, Oct. 12, 1960.

\2 Orr, W. I., and Cowan, S. D., **All About Cubical Quad Antennas**, 3rd edition, (Wilton, CT: Radio Publications, Inc., 1982), p. 6.

\3 Woram, J. M., "Broadcasting and Recording in Ecuador," **dB Sound Engineering**, April 1982, pp. cover, 47.

\4 Doherty, R. M., "Cubical Quad Antennas With Spreader-Reinforced Crossarms," U.S. Patent No., 4,138,682, Feb. 6, 1979.

\5 See note 2, p. 33.

\6 "Pointers on the Gem Quad Antenna," **QST**, Feb. 1979, p. 50, and "Product Review," **QST**, Jan. 1978, p. 38.

\7 Partridge, E. G., "Antenna Wire Support Structure," U.S. Patent No. 3,532,315, Oct. 6, 1970.

\8 "Fiberglass Poles for Antenna Construction," Hints and Kinks, **QST**, March 1983, p. 41.

\9 See note 2, p. 30.

\10 "FCC Takes Big Step Toward Putting WARC-79 Into U.S. Law," **QST**, March 1983, p. 56.

\11 Hall, G., ed., **The ARRL Antenna Book**, 14th edition (Newington: ARRL, 1982), p. 9-7.

\12 See note 2, pp. 40, 46, 52.

\13 Eckersley, R. J., ed., "Working DX Stations," **Amateur Radio Operating Manual**, 2nd edition, (London: RSGB, 1982), p. 36.

\14 See note 2, p. 89 and note 11, p. 6-16.

\15 Orr, W. I., and Cowan, S. D., **All About Cubical Quad Antennas**, 2nd edition, (Wilton, CT: Radio Publications, Inc., 1970), p. 40.

\16 See note 11, p. 6-23.

\17 Lindsay, J. E., "Quads and Yagis," **QST**, May 1968, pp. 11-19, 150.

\18 Bergren, L., "The Multielement Quad," **QST**, May 1963, pp. 11-16.

\19 See note 2, p. 50 and note 6, p. 50.

\20 See note 2, pp. 60-74 and note 11, pp. 5-14, 9-9.

\21 See note 2, pp. 75-80, 107.

\22 See note 11, pp. 8-21, 9-5.

\23 See note 11, p. 9-8.

\24 See note 11, pp. 8-22, 8-24.

\25 Woodward, G., ed., **The Radio Amateur's Handbook**, 60th edition, (Newington:ARRL, 1983), p. 17-6.

\26 See note 11, p. 8-23.

\27 **Structural Standards for Steel Antenna Towers and Antenna Supporting Structures**, EIA Standard RS-222-C, Electronic Industries Assoc., March 1976, available from EIA, 2001 Eye St., N.W., Washington, D.C. 20006.

\28 See note 11, p. 9-19.

\29 See note 11, p. 8-29.

\30 Ellis, C. J., "A Novel Way to Mount a Rotary-Beam Antenna," **QST**, May 1979, pp. 32-33.

\31 See note 11, p. 9-24.

\32 See note 2, p. 37.

\33 See note 11, pp. 3-12, 3-14, 3-20.

\34 Gibilisco, S., "What Does Your SWR Cost You?," **QST**, January 1979, pp. 19-20.

\35 See note 11, p. 5-22.

\36 See note 11, pp. 6-16, 6-18.

Table 1

Band (meters)	f (MHz)	S_D (ft)	S.A.*$_D$ (ft)	S_R (ft)	S.A.*$_R$ (ft)	Spacing (ft)
20	14.17	17.57	12.90	18.14	13.32	8.33
17	18.12	13.74	10.05	14.18	10.38	6.51
15	21.22	11.73	8.56	12.11	8.84	5.56
12	24.94	9.98	7.25	10.30	7.49	4.73
10	28.85	8.63	6.25	8.91	6.46	4.09

All length dimensions are feet

f = design resonant frequency in MHz

S_D = 249/f = driver side length

S_R = 257/f = reflector side length

S.A. = spreader attach length*

S.A. = [side/(2sin45° x cos18°)] - hub clearance

S.A. = 0.7435 x side - 0.167

Free space wave = 984/f

Spacing = 0.12 wavelength = 118/f

*Measured from hub end of spreader

Bicycle Wheel Quad

By Dave Guimont,* WB6LLO

*5030 July St., San Diego, CA 92110

The prototype of this antenna was originally developed for OSCAR 8 uplink modes A and J. Performance in those modes was highly satisfactory, and uplink comparison checks were made against a 16-element crossed Yagi operating in the following configurations: vertical, horizontal, axial, LHCP and RHCP. The Bicycle Wheel Quad signals equaled or bettered those configurations 90% of the time. The "bicycle wheel" construction is not my design, but is patterned after an article in **Ham Radio**.[1]

Aluminum clothesline wire (no. 9), drilled at the spoke junctions, was used in the original antenna. Two years of exposure to a mild, but corrosive, atmosphere finally took its toll. The 1/4-inch aluminum tubing and split-ring anchoring method used here should prove more durable (Figs. 1 through 3).

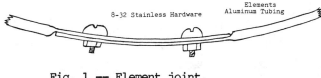

Fig. 1 — Element joint.

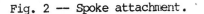

Fig. 2 — Spoke attachment.

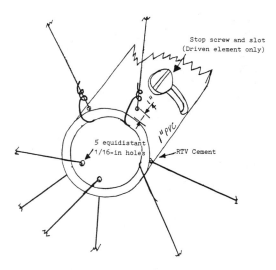

Fig. 3 — Hub detail.

The materials are inexpensive. Ordinary hand tools will do the job. One feed line provides vertical, horizontal or any degree of polarization in between. The turning radius is less than 30 inches, and my antenna weighs 2 pounds, 12 ounces.

Various electro-mechanical polarity rotation systems were considered. The fishline lanyard works quickly and efficiently; other devices may be required at your location. Fig. 4 shows the rotation arm support used in my setup.

Gain figures are unknown. Again, on-the-air performance comparisons are the only methods available to me. Brief testing on OSCAR 10 indicates reception is equal to or better than a 6-turn RHCP helix or a 16-element crossed Yagi in all configurations. Element spacing was determined empirically. Prototype spacings (no. 9 wire) were within 1/2 inch of these dimensions (Fig. 5).

A field strength meter and SWR bridge were used to set the elements and gamma for maximum gain. Minimum SWR and maximum gain did not coincide. These dimensions provided maximum gain at a SWR of 1.3:1 measured at the antenna. Photos and the five illustrations provide construction details.

Mount the antenna so a pull on the fishline provides vertical polarization (feed point at the side). Rig a small diameter bungee so its tension plus the weight of the coaxial feed line returns the feed point to the bottom (horizontal polarization).

My Bicycle Wheel Quad is shown in Fig. 6. I would be interested in comments and improvements from builders. Information regarding actual gain would be particularly appreciated. CU on OSCAR 10!

References

[1] Kennicott, J. W., W4OVO, "Dream Beam," **Ham Radio**, p. 12, May 1980.

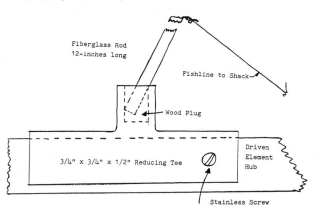

Fig. 4 — Rotation arm support.

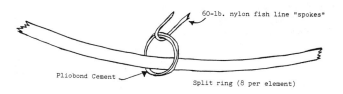

Fig. 5 — Element spacing.

MATERIALS

QTY	
1	4 ft, 6 inch length 1/2-inch schedule 40 PVC (boom)
4	6 inch length 1 inch schedule 40 PVC (element hubs)
30 ft	1/4-inch aluminum tubing
80 ft	60-lb test nylon fishline (spokes)
32	3/8-inch split rings (perimeter spoke supports)
1	3/4 inch x 3/4 inch x 1/2 inch T schedule 40 PVC (rotation arm support)
1	1 ft x 3/8 inch fiberglass rod (rotation arm)

GAMMA MATCH

BNC connector
20-pF variable capacitor
6 inch, no. 10 copper wire
3/8 inch x 2 inch stainless strap
8-32 x 1/2 inch stainless hardware

ELEMENT DIMENSIONS

	Diameter	Circumference
Reflector	27 inch	84 3/4 inch
Driven	25 3/4 inch	80 7/8 inch
1st Director	24 13/16 inch	78 inches
2nd Director	23 7/8 inch	75 inches

(I added 6 inches to the circumference for overlap, and flattened and secured it with 8-32 stainless hardware.)

SPACING

Reflector to driven	15 1/2 inch
Driven to 1st director	13 1/2 inch
1st director to 2nd director	18 1/2 inches

Fig. 6 -- A view of the completed Bicycle Wheel Quad.

Quad Antennas for 80 and 160 Meters

By William M. Kelsey,* N8ET

*2716 C.R. 26, Mt. Cory, OH 45868

During 1977 and 1978, I constructed and used a 3-element quad antenna for 40 meters.[1] It was the best 40-meter antenna I ever used. European stations could be heard over most of the east coast stations, and a simple switching unit, constructed with the quad, enabled the quad's direction to be reversed so I could work ZL/VK stations.

I wanted to try the same arrangement on 80 meters, but was transferred to England for five years. It was difficult to erect even a small tribander where I lived, so the 80-meter quad was out of the question. Returning to Ohio in 1983, I again had room to construct the 80-meter quad antenna and found that KS8S had just erected a 180-ft tower. I began to think about a quad for 160 meters, also.

General Design

A **QST** article I was reading at the time indicated that a quad-loop reflector could be created by cutting the loop the same size as the driven loop, and adding approximately 150 ohms of inductive reactance.[2] A director could be constructed in the same manner, except that approximately 150 ohms of capacitive, not inductive, reactance was to be added. See Fig. 1.

By applying some basic transmission line theory and making use of a Smith Chart, 150 ohms of reactance can be created at one end of the feed line. This is done by placing a capacitor at the other end of the line. The line length transforms any capacitive reactance at one end to 150 ohms of capacitive reactance at the other end; or, if the line is a bit longer, to 150 ohms of inductive reactance. In fact, any value of reactance can be created using the correct line length.

As an example, Fig. 2 shows how 150 ohms of inductive reactance can be created by placing a capacitor of 900 ohms reactance at the end of a length of a 450-ohm line, 0.226 wavelength long.

This is the distance shown on the outer edge of the Smith Chart between point A (150-ohms inductive) and point B (900-ohms capacitive). Note that if 300-ohm line were used, the line length would be different. Fig. 3 shows how a quad reflector would be constructed using 900 ohms of capacitive reactance and a length of 450 line 0.226 wavelength long. For an additional explanation of how the Smith Chart works and how to use it, see **The ARRL Antenna Book.**[3]

By taking the above ideas one step further, and using a fixed length of line, either a director or a reflector can be created by changing the value of capacitance at the end of the line. For example, if 0.16 wavelength of 450-ohm line were used between the loop and the capacitor, the loop would act as a reflector, if the capacitor were 373 ohms (point A, Fig. 4). If the capacitor were 1755 ohms (point B, Fig. 4), it would act as a director. This can be done by switching the two capacitors at the end of the line, or by adjusting the value of a variable capacitor. Study the Smith Chart for a few minutes. You will also note that with an appropriate length of open (or shorted) line you can create either a director or reflector from a one-wavelength loop.

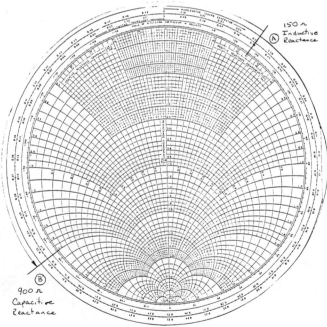

Fig. 2 — By placing a capacitor at one end of your feed line, an inductive reactance can be created at the other end of the line.

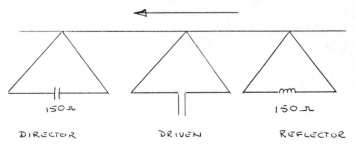

Fig. 1 — A basic quad with 150-ohm capacitive reactance added to the director.

Transmission Lines

By performing a few hypothetical design problems on the Smith Chart (transforming reactances along a transmission line), and using a higher impedance line, it will be much easier to create a reflector and director by switching the two capacitors and using a fixed length of line. Because of this, and because a loop is balanced, I have always used open-wire line between the loop and the capacitors. Open wire is available commercially, or can be constructed at home as shown in **The ARRL Antenna Book.** The characteristic impedance of the open wire depends primarily on the wire size used and the spacing between the wires. Again, see **The ARRL Antenna Book** for information.\4

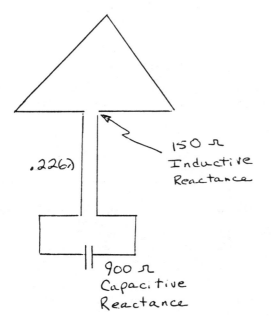

Fig. 3 — This shows how a quad reflector would be constructed using 900 ohms of capacitive reactance and a length of 450 line 0.226 wavelength long.

I constructed my open wire two different ways. The first method I tried used thin wall PVC tubing. This resulted in a rather heavy line which was difficult to handle and the spacers did not stay on as well as I had hoped. I would **not** recommend this method of constructing open-wire line.

The second method I tried was much more successful. I cut several plastic milk bottles into strips about 1 inch x 4 inches. Using an awl, I poked two holes in each end of each strip. The wire was fed through both holes, and the plastic was springy enough that the spacers maintained their position on the line (see Fig. 5). Amazingly, this open-wire line survived several Ohio blizzards. As a side benefit, I found that triangular pieces of the same plastic milk bottles made excellent strain-relief insulators for the corners of the quad elements, and I constructed an excellent center insulator out of the same material. This line was easier to construct and is light; the plastic can be cut with scissors. It now seems as if I have a never-ending supply of spacer material.

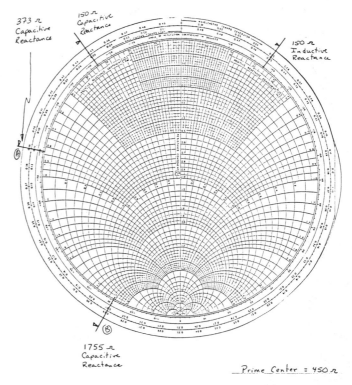

Fig. 4 — Using a fixed length of line and by changing the value of capacitance at the end of the line, a director or reflector can be created.

Antenna Dimensions

Spacing of the quad elements is not critical. **The ARRL Antenna Book** states that spacings of 0.14 to 0.2 wavelength are common. I used about a 40 ft spacing for a 2-element 80-meter quad because I had a 40-ft boom on hand to take the elements from. Spacings between 25 ft and 65 ft should be acceptable.

Loop dimensions are calculated using the standard formula for the driven element of a quad:

$$l = \frac{1005}{f}$$

where

l = length in feet
f = frequency in megahertz

Loop dimensions for various frequencies in the 80- and 160-meter bands are shown in Tables 1 and 2, respectively.

Practical Design Example - 80 Meters

A 3-element quad for 3.5 MHz could be constructed as follows:

Loop size from Table 1: 287 ft for each of the three elements.

Line length: A line length of 0.14 wavelength of 450-ohm line was chosen to go between the parasitic loops and the capacitors. This allows capacitors of a reasonable value to be used at the ends of the line to make either a director or reflector. Using the formula below, 0.14 wavelength was calculated to be approximately 38 ft of line:

$$l = \frac{nk \ (984)}{f}$$

where
    l = length in feet,
    n = number of wavelengths,
    k = velocity factor of the line
        (about 0.96 for open-wire line)
f = frequency in MHz.

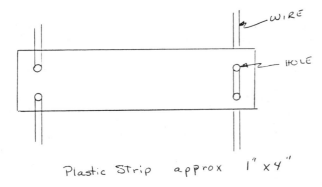

Fig. 5 — Plastic strips for use as spacers were cut from milk bottles. With the help of an awl, two holes were poked in each end of each strip to allow the wire to be fed through. The plastic is springy and light enough that the spacers maintain their position on the line.

Capacitor value - Reflector: The reactance required at the end of the 0.14-wavelength 450-ohm line, to make a reflector, is approximately 293 ohms. This was determined by rotating the Smith Chart 0.14 wavelength toward the load from the 150 ohm (inductive) point, and reading the reactance from the chart. The capacitance is calculated using the formula:

$$C = \frac{1}{2\pi f X c}$$

where
    C = capacitance in farads,
    Xc = capacitive reactance in ohms,
    f = frequency in hertz.

The capacitor should, therefore, be about 155 pF. A variable capacitor of about 250 pF maximum will work nicely.

Capacitor value - Director: To make a director, the capacitor required at the end of the 0.14 wavelength of 450-ohm line is approximately 1125 ohms. This works out to be about 40 pF. A 100 pF, or another 250 pF, variable would work well here.

A diagram of this quad is shown in Fig. 6. The quad could be suspended from a large boom, or a line between two tall supports. The quad I built had the apex at about 90 ft and the bottom was about 20 ft above the ground. The corners were pulled out and supported by small nylon cord. Element shape is not critical. The direction the main lobe is headed is changed by switching the relays shown in Fig. 6.

Practical Design Example - 160 Meters

The 160-meter quad could be designed in the same manner as the 80-meter quad, however I only had

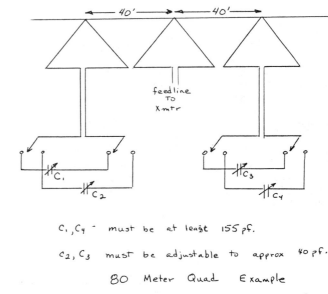

C_1, C_4 - must be at least 155 pf.

C_2, C_3 must be adjustable to approx 40 pf.

80 Meter Quad Example

Fig. 6 — A three-element quad for 3.5 MHz is constructed using the capacitor values listed. C1 and C4 form the reflector, while the director consists of C2 and C3.

one support available which was high enough for a 160 quad. The 160 quad was therefore set up using only one switchable parasitic element. The peaks of both elements were hung from the same support and the bottoms were spread out as far as was physically possible.\5 The top was at 170 ft and the bottom was about 40 ft above the ground.

Dimensions for a quad designed for 1.8 MHz using 0.14 wavelength line are as follows.

Loop size: 558 ft
Line length: 73 ft
Capacitor size - reflector: pF (450-pF variable)
Capacitor size - director: 79 pF (150-pF variable)

Tuning the Quad

The elements should be in place before performing tuning adjustments. The parasitics will affect the resonant frequency of the driven element. First, adjust the resonant frequency of the driven element using an SWR or noise bridge, and trim the length of the driven element. Then adjust the parasitic elements by adjusting the capacitors, using a receiver for the best rejection of signals off the back of the antenna. This will correspond closely to the point of maximum forward gain. It is much easier to see a dip in signal strength off the back than to adjust for a peak in a signal off the front. The peak is very broad, while the dip off the back is sharper. Use a horizontally polarized signal when doing these adjustments, preferably a long distance away. If that cannot be done, use a horizontal dipole several wavelengths away to provide a signal. **Do not** use a vertical or a short-wire antenna as a source or you will receive strange results.

Final Comments

I hope this article has provided ideas and enough information to encourage other amateurs to build and design a few monster quads. The article presents very few specific dimensions because it would be difficult to reproduce the quads exactly as

I built them. In any event, there are several variables that can be adjusted to fit a particular physical situation; spacing, height, line length, and capacitor value are just a few examples. Note that the lines can be extended by adding multiples of half wavelengths of line, and the capacitors could be brought into the operating position. The capacitors on my quads have all been mounted in plastic ice cream containers; they were receiving type capacitors, and the relays were small DPDT types. The driven element was fed directly with 52-ohm coaxial cable.

I would be happy to correspond with anyone having questions about these quad antennas.

References

\1 Kelsey, William M., "A Three-Element Switchable Quad for 40 Meters," **Ham Radio**, p. 26, Oct. 1980.

\2 Kaufmann, John E., and Kopec, Gary E., "A Convenient Stub-Tuning System for Quad Antennas," **QST**, p. 18, May 1975.

\3 **The ARRL Antenna Book,** 14th edition, Chapter 3.

\4 See Reference 3.

\5 Pfaff, William, "The K2GNC Giza Beam," **Ham Radio**, p. 52, May 1981.

Table 1. Loop Lengths — 80 Meters

FREQ (MHz)	LOOP LENGTH (FT)
3.5	287
3.6	279
3.7	272
3.8	265
3.9	258
4.0	251

Table 2. Loop Lengths — 160 Meters

FREQ (MHz)	LOOP LENGTH (FT)
1.800	558
1.850	543
1.900	529
1.950	515
2.000	503

Gossamer Quad Update

By R. F. Thompson,* W3ODJ/ZF2CD

*Highway 5, Box 31, Waldorf, MD 20601

It takes less than you think to keep quad loops in the air. This feather weight 20-meter beam has been flying for four years.

What is a Gossamer?

The gossamer quad is electrically similar to most conventional two-element 20-meter quad antennas. The essential difference between gossamer and conventional construction is in the mechanical support and steering structures that keep the one-pound radiating and reflecting loops in the air, properly spaced, spread and oriented. All the parts of a conventional quad are pushed up from below by a massive, compressive structure that includes an expensive metal tower. The essential parts of the Gossamer quad, however, hang from an overhead tensile support system that is both lightweight and inexpensive.

Fig. 1 is a perspective sketch of the gossamer quad, and Figs. 2 through 5 magnify some of the assembly details. The "sky line" suspension cable is simply a nylon rope stretched tautly over the crowns of two tall trees, and it supports the two quad loops and their lightweight framework. The mechanical parts of the antenna, such as the rotator, spindle and steering pole, sit atop a ground based vertical cedar pole. There is nothing new in dividing the load between antenna parts that will be high above the ground, and those parts that are lower in height. It has been tried and proved at Arecibo in Puerto Rico where the largest radio telescope in the world uses suspension cables to support its antenna high above a large reflector resting on the ground.

The framework for the two gossamer quad loops consists only of four spacer poles and one spreader pole, and weighs about eleven pounds. The supporting ropes weigh less than ten pounds, so only about 20 pounds of material are needed to keep the one-pound loops near tree top height and properly positioned. Conventional quads use several hundred pounds of tower, mast, boom, spiders and spreaders to do the same thing. Towers do have one advantage, however. They can raise antennas higher than natural supports.

The gossamer quad takes advantage of the overhead support that the sky line provides, by utilizing the tension in the loop wires to supply most of the support for the loops and their framework. Thus, the loop wires are not only the most important electrical component of the antenna, they are very important mechanical components as well. In short, the essential difference between conventional and gossamer construction is that conventional quads use massive, compressive support, while the gossamer uses lightweight, mostly tensile support.

The First Four Years

The original gossamer quad was described in QST.\1 With some minor modifications on my part, it has been actively used for over four years at my

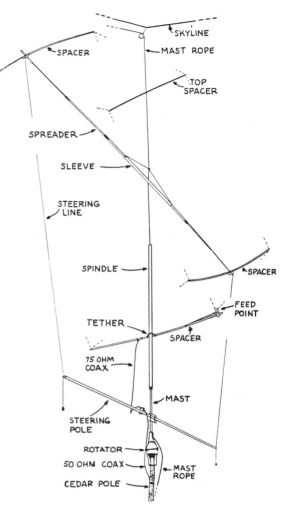

Fig. 1 — A perspective sketch of the support system looking up from the ground. The sky line, pulley and mast rope support the spacers, spreader, loops and the 75-ohm matching section. The corners of the loop wires are shown as dashed lines at the ends of the four spacer poles. The cedar pole supports the rotator motor, steering pole, spindle, and the 50-ohm feed line. The steering lines are tied to the ends of the spreader.

location. Only one important weak spot has been found in the original design, and this has been strengthened considerably by a simple fix described below. In addition, the original lower rigging has been replaced by a self-adjusting rigging that uses a spindle pole to keep the bottom of the antenna centered over the rotator motor. Neither of these improvements add any significant weight to the load on the sky line.

The original top rigging on the gossamer quad was proved to be highly satisfactory. The sky line, pulley and mast rope easily survive the weather's worst tricks. As mentioned, the sky line is simply a nylon suspension rope, and the pulley "sky hook" is fixed on the sky line directly above the rotator motor. The quad loops and their framework hang from the mast rope which passes up through the pulley and

The rotatable gossamer quad is a two-element 20-meter beam which has a minimum of heavy, self-supporting components. The loops are mechanical parts of a lightweight framework that hangs from an overhead line.

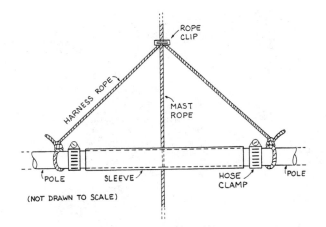

Fig. 2 -- Spreader-sleeve assembly. The two halves of the spreader pole are joined by slipping the large ends inside the spreader sleeve. Hose clamps prevent either half from pushing the other half out of the sleeve. The assembly hangs by the harness rope from a metal rope clip on the mast rope. The mast rope passes freely behind the sleeve.

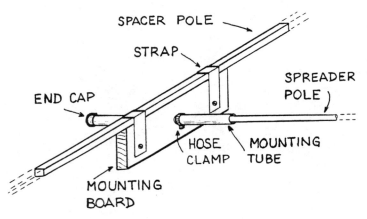

Fig. 3 -- Spacer-spreader mounting. The spacer pole is fastened to the mounting board by two U-shaped aluminum straps which are bolted to the board. The mounting tube passes through a snug hole in the board. The small end of the spreader pole is inserted into the tube and pushes against the end cap. The hose clamp transfers the push to the board and spacer, and this spreads the loops into a diamond shape.

then goes over the crown of a nearby tree and down to a cleat where it is secured after it is used to raise the quad loops to operating height, or to lower them for repairs. During the summer of 1982 the mast rope was sabotaged by squirrels! They apparently enjoyed chewing through the nylon strands. Luckily, the damage was spotted before the rope broke, and the badly frayed portion was cut out. The spliced rope was then rerouted away from the squirrel's playground.

The framework for the loops, consisting of the spreader and spacer poles, has also been satisfactory. The 27-ft spreader pole is the only compressive part of the framework, and it weighs only 2.6 pounds. The two halves of the spreader pole were originally aligned and held together by thin aluminum L-angle stock and hose clamps. Unfortunately, this design subjected the thin L-angle to forces

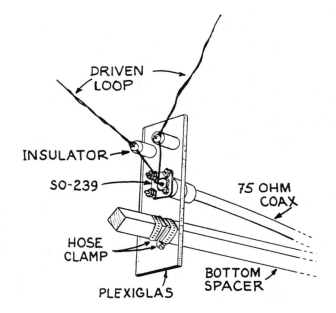

Fig. 4 -- Feed-point assembly. The ends of the driven loop are attached to short standoff insulators mounted on a transparent Plexiglas plate. Short wires connect the loop to an SO-239 coaxial cable jack. A square hole in the Plexiglas provides a snug fit for the bottom spacer, and the plate is held in place between two hose clamps attached on the spacer.

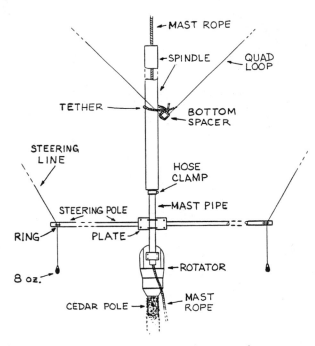

Fig. 5 -- Lower support structure. The mast rope passes through the spindle and exits the bottom of the mast pipe. The bottom spacer is tethered to the spindle so that it is always centered above the rotator, and it is free to move vertically. The spindle slips down over the mast pipe and rests on a hose clamp tightened around the mast pipe. U-bolts through the mounting plate attach the steering pole to the mast. The steering lines pass through metal rings hose clamped to the ends of the steering pole, and are kept taut by 8-ounce weights.

which easily distorted the aluminum, and in April 1982 the L-angle ruptured during a week-long attack of severe winds.

The torsional weakness of the L-angle has been the only significant design flaw discovered during more than four years of continuous service. The L-angle has been replaced by a round aluminum tube, the spreader sleeve (Fig. 2). The tube is a half-pound heavier, but much stronger structurally.

The original lower rigging used four guy lines to limit motion of the lower spacer and keep the spacer in position above the rotator motor. The guy ropes, however, needed to be adjusted during times when the rain or ice changed the height of the sky line. When the guys were too loose, the lower spacer was poorly constrained; when they were too tight, they would pull down on the sky line and lower the height of the antenna. The original rigging has been replaced by a new design that is self adjusting, and it does not add any load on the sky line. The lower rigging keeps the bottom part of the antenna exactly centered on a spindle pole which slips over the mast pipe mounted on the rotator motor. A loose loop of nylon rope tethers the center of the spacer to the spindle and allows the spacer to ride up and down the spindle whenever changes in the weather change the height of the sky line. After more than a year in service, it seems that little could be done to improve on this design.

Parts and Materials

Table 1 catalogs the major components of the gossamer quad in two categories: parts supported by the sky line, and parts sitting atop the cedar pole. Most of the material was purchased from local electronics, lumber and hardware stores, but the two halves of the spreader pole were obtained through mail order from Skylane Products.\2 These tough, but light weight poles are called Dynaflex fiberglass spreaders. Each consists of three telescoping sections that extend to a maximum length of 15 ft. These two poles are joined in the spreader sleeve and the lengths of the sections are adjusted and locked by small hose clamps to give the spreader pole a total length of 27 ft.

The original sky line is still in service and shows no sign of wear. It is made of 0.25-inch-diameter Goldline nylon rope purchased from REI.\3 REI has discontinued selling Goldline in 1982, but they now offer a superior rope called Skyline II. A 1984 catalog lists 1/4-inch-diameter Skyline II at 15 cents per foot, obtainable in any length up to 600 feet. Compared to Goldline, the new Skyline II is 100% more abrasion resistant when dry, 50% more abrasion resistant when wet, and 20% more resistant to the degrading effects of ultraviolet rays.

Any rigid pole having a length of 13 feet or more can be used for the steering pole. The gossamer quad uses a fiberglass pole that has a honeycomb cellular cross section. This pole is a four-pound spreader designed for use in conventional quads.(A conventional quad would use eight of these, 32 pounds of spreader in all. The gossamer uses less than three pounds.)

The cedar pole was once a tree in my back yard, and it is the only eye-level object in the support system, blending naturally with its surrounding. It elevates the rotator about 13 ft above ground, and supports the weight of the rotator, steering pole, and spindle pole. Since the spindle pole is a fair guide for the lower mast rope, the slender cedar pole also absorbs some of the shocks of wind loading on the gossamer quad. The original cedar pole is still being used, and shows no sign of weakening. Untreated cedar fence posts have a life expectancy of between 15 and 20 years.\4

Conclusion

This article presents observations made during the gossamer quad's first four years of operation, and it has given details of changes made since the first gossamer quad article was published in **QST**. The **QST** article contains additional information not repeated here, which describes how to attach the loops to the spreaders, how to launch the sky line using a sling shot, formulas for the lengths of the quad loops, and use of a quarter-wave transformer to feed the driven loop.

Other articles containing useful ideas for gossamer construction have been written. VE3CUI described his gossamer Yagi in **CQ** and KC7M described some useful knots in a recent issue of 73.\5,6

Finally, the caveat must be repeated that the gossamer quad is not a rigid structure, and it may be a poor design for use in locations that are subject to frequent strong winds. But the design has proved itself to be rugged, durable and well suited to the usually calm weather in Maryland. In my backyard, a many-hundred-pound metal tower is not needed to keep one-pound loops near tree-top height.

Acknowledgement: I would like to thank Les Salter, KA3DCW, who assisted me with various computer tasks such as transferring files from my Apple II floppy disk to an IBM PC diskette.

References

\1 Thompson, R. F., W3ODJ, "Build A Gossamer Quad," **QST**, December 1981.

\2 Skylane Products, 359 Glenwood Ave., Satellite Beach, Florida, 32937 (305-773-1342).

\3 REI, P. O. Box C-88125, Seattle, WA 98188.

\4 **Back To Basics**, Reader's Digest Association, 1981.

\5 Swynar, E. P., VE3CUI,"A Tree-Mounted Vertical Yagi Array," **CQ**, August 1982.

\6 Jacobs, G., "All Tied Up In Knots?" **73**, Oct. 1982.

TABLE 1. GOSSAMER QUAD COMPONENTS

COMPONENTS SUPPORTED BY THE SKY LINE

COMPONENT	MATERIAL	DESCRIPTION	SIZE *	WEIGHT pounds
TOP SPACER	WOOD	ROUND CLOSET POLE	8 ft 7 in LONG 1.25 in DIA	3.0
LEFT SPACER		SQUARE THIN POLE	8 ft 7 in LONG 0.75 in SQUARE	1.0
RIGHT SPACER	WOOD			1.0
BOTTOM SPACER				1.0
SPREADER POLE	FIBERGLAS	ROUND POLE TELESCOPING 6 SECTIONS	27 ft LONG Max 1.06 in DIA Min 0.50 in DIA	2.6
SLEEVE	ALUMINUM	ROUND TUBING	6ft LONG 1.25 in OD 1.12 in ID	1.3
SPACER/SPREADER MOUNTING BOARDS	WOOD	RECTANGULAR (2 needed)	6 in x 3 in 0.75 in THICK	0.25 0.25
MOUNTING TUBES	COPPER	ROUND PIPE (2 needed)	10 in LONG 0.63 in OD	0.25 0.25
FEED POINT	PLEXIGLAS (clear)	RECTANGULAR PLATE	6 in x 2.25 in 0.25 in THICK	0.13
DRIVEN LOOP	COPPER	SQUARE LOOPS	17.9 ft/SIDE	1.0
REFLECTOR	WIRE	14AWG 7-strand	18.4 ft/SIDE	1.0
QUARTER WAVE TRANSFORMER	COAXIAL CABLE	RG-11/U 75 ohm	Approx. 11.6 ft (Tuned to 14 MHz)	1.0

COMPONENTS SUPPORTED BY THE ROTATOR

COMPONENT	MATERIAL	DESCRIPTION	SIZE *
SPINDLE	PVC	ROUND PIPE	10 ft LONG 1.88 in OD, 1.56 in ID
STEERING POLE	FIBERGLAS	ROUND POLE VERY RIGID	13 ft LONG 1.12 in DIA
MOUNTING PLATE	ALUMINUM	RECTANGULAR PLATE	7 in x 4.5 in 0.125 in THICK
MAST	ALUMINUM	ROUND PIPE	6 ft LONG 1.5 in OD, 1.25 in ID

(in=inch=2.54 cm, ft=foot=0.3048 m)
(DIA=Diameter, OD=Outer DIA, ID=Inner DIA)

The PV4 Quad — A New Twist

By John DeWitt,* AI9P

*21 Central Ave., Haledon, NJ 07508

A new twist to the conventional tri-band 3-element quad? The PV4-Quad adds a 40-m helix Yagi to the usual 20-, 15-, and 10-m quad wire loops. This is a homemade antenna project with all the basic exercises:

A. Gathering concepts, ideas and resources
B. Construction
C. Tuning, testing and evaluation
D. DX enjoying

Background

The HW-40 helical beam described in February 1974 **QST,** was my selection for a small size beam. A friend convinced me to wind the helix with copper tape rather than wire. About the time I was to start its construction, quad fever struck! Several evenings of wrapping and unwrapping the test helix windings derived a 7.2-MHz dipole only 23 feet long. The spreaders of a 20-m quad happen to be 24 feet long! Ad hoc viritus...Why not wind the 40-m helix on the horizontal spreaders of a quad in a diamond configuration? Hence, the PV4-Quad is a 4-band quad built of PVC pipe.

Concepts, Ideas and Resources

I was surprised that the PVC "schedule 40" tubing, used in the project, weathered six Chicago winters. Hopefully, the "tensegrity" design, which flopped around without breaking in the Windy City, will hold up during New Jersey hurricanes. Tensegrity is a contraction of tension with integrity. It means that something is just limber enough to flap, but not stiff enough to snap.

Tri-gamma matching elements combined with an open wire 40-m 1/4-wave matching section allow the array to be driven by a single coaxial feed line. This article covers my experience from Phase I of 1977 through Phase III in 1984. It contains construction and tuning details, accompanied by photos and tabulated sizes and spacing of the elements.

Construction materials include 500 feet of no. 14 gauge stranded copper house wiring, telescoped aluminum vertical spreaders, boom and horizontal spreaders of PVC "schedule 40" tubing, and a unique 40-meter helix wound of copper tape. Fig. 1 presents the overall view of the array. Tables following the text provide dimensions for wire loops, helix spacing, and tri-gamma match elements.

The 40-m helix antenna consists of 1/3 inch wide tape windings on the horizontal spreaders of each element. The driven element tape is copper, while the director and reflector are wound of aluminum tape. Thus, we have a 3-element helix Yagi.

A hairpin matching loop made of no. 8 gauge copper rod is tuned to match the 450-ohm open wire, 31.5 foot, 1/4-wave matching line from the tri-gamma feed point to the 40-m helix driven element 15-ohm feed point. The tri-gamma match feed point for RG-8 50-ohm coaxial cable is tuned by the three individual gamma match capacitors, and a fourth compensator capacitor across the open wire line between the 20-meter and 10-meter quad feed points.

The project is a challenge and is time consuming. Resources for the design were obtained from local plumbing supply houses, and the articles noted in the list of references. The various sizes of aluminum tubing were found in a do-it-yourself home center. PVC "schedule 40" pipe is available in many sizes at most plumbing stores. The 500-ft reel of house wire was purchased from an electrical supply store. See the materials list for more parts resource information. The most durable resource is my wife, Dolores, who perfected procedures to cut and wind the helix tapes.

Construction

Fig. 2 and Table I present details of the tri-gamma match elements, originated by Lew McCoy, W1ICP. Each of the three gamma matching elements were individually tailored per my notes in Table I. Spacing to the wire loop, as well as distance from the loop apex to the tap point, was derived with the tuning capacitor set at midrange. The finishing touches were then added with a series of SWR adjustments using the four capacitors. See the section on tuning for more details.

Table II lists the dimensions and tuning frequencies for the wire loops. Note the higher frequency for the director and the lower frequency for the reflector. All quad loop elements are closed loops, so no tuning stubs are used.

The helix windings for 40 meters were tuned with a dip meter coupled to the end of the helix. A tape width of 1/3 inch with 1/3 inch spacing between turns approximates 1-1/2 turns per inch, on a 1-11/4 inch OD tube. After several trial windings, a pitch of 14 turns per foot for a total length of 24 feet tuned to 6.8 MHz. I was able to tune across the 40-m band from 6.5 to 7.5 MHz by changing the 24-ft winding by 2-1/2 feet. Table III shows some experimental helix winding data.

A note of precaution: Apply the metal tapes to the PVC surface using a light touch. Do **not** stretch the tape! Allow some slack in the windings. A loose winding permits bending the PVC spreader without snapping the tape.

There were a few other factors that changed the helix tuning frequency: Varnish (one coat) lowered the frequency; the coupling of the 20-, 15-, and 10-meter loops detuned the helix and also changed the loop's tuning. Height above ground caused a 100-kHz shift higher at 30 feet compared to testing at 8 feet in a horizontal attitude.

Figs. 3 through 6 show the boom and element construction features. The boom hub for each element is "laminated" of a wood clothes pole core, and an intermediate 1-1/2-inch ID PVC tubing, surrounded by

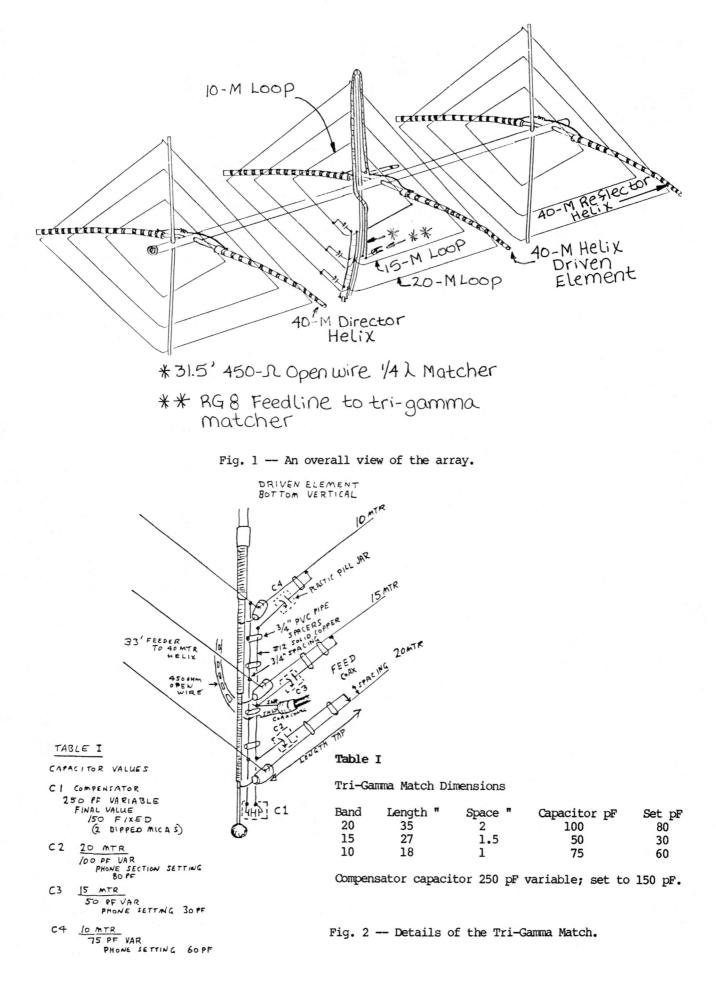

10-M LOOP

40-M Reflector Helix

40-M Helix Driven Element

40-M Director Helix

15-M LOOP

20-M LOOP

✳ 31.5' 450-Ω Open wire ¼ λ Matcher

✳✳ RG8 Feedline to tri-gamma matcher

Fig. 1 — An overall view of the array.

DRIVEN ELEMENT BOTTOM VERTICAL

10 MTR

15 MTR

20 MTR

C4 — PLASTIC PILL JAR

¾" PVC PIPE SPACERS

#12 SOLID COPPER

¾" SPACING

FEED COAX

SPACING

33' FEEDER TO 40 MTR HELIX

450 OHM OPEN WIRE

C3

COAX COVER

C2

LENGTH TAP

C1

TABLE I

CAPACITOR VALUES

C1 COMPENSATOR
250 PF VARIABLE
FINAL VALUE
150 FIXED
(2 DIPPED MICAS)

C2 20 MTR
100 PF VAR
PHONE SECTION SETTING 80 PF

C3 15 MTR
50 PF VAR
PHONE SETTING 30 PF

C4 10 MTR
75 PF VAR
PHONE SETTING 60 PF

Table I

Tri-Gamma Match Dimensions

Band	Length "	Space "	Capacitor pF	Set pF
20	35	2	100	80
15	27	1.5	50	30
10	18	1	75	60

Compensator capacitor 250 pF variable; set to 150 pF.

Fig. 2 — Details of the Tri-Gamma Match.

a larger 2-inch ID PVC tube. The sections of the boom between the center of the driven element and the reflector or director are hollow 2-inch ID tubes. Aluminum mesh around the spreaders acts as a cushion for the boom laminate. The mesh allows for some floppy movement of the horizontal and vertical spreaders.

Note that the wooden core of the hub sections must be varnished completely before encasing it with the PVC tubes. Stainless steel screws hold the spreaders in place at the hubs.

A cork washer, fabricated from linoleum, cushions the boom structure where the element hubs meet the hollow boom tubes. The hub section that telescopes inside the boom tube may be 3 or 4 feet long. This provides a "handle" for carrying the fully assembled element around the yard during tuning tests. Fig. 7 shows a view of a typical hub.

Vertical spreaders of telescoping aluminum are 1 inch OD at the boom, then 7/8 inch OD, with 3/4 inch OD at the ends. Lightning protection is provided by a ground strap system from each vertical spreader, across the boom and down the tower to ground.

Unique PVC tubing wire holders are attached to the spreaders by galvanized no. 16 gauge straps. The wire loops are firmly anchored by a 3-step lock-in loop through the PVC holder (Fig. 8). Dress the wire through the wire holders such that the wire is clear of the aluminum vertical spacers. Six years of Chicago weather didn't wear through the house wire insulation. Note that the wire holders on the bottom vertical spreader of each element must be movable to provide slack in the wire loops for tuning.

In Phase III, short nylon string holders are used along the horizontal spreaders for less wear and tear on the helix tape windings. Each is protected against abrasion by a layer of 3M "strapping" tape on top of the copper or aluminum tape winding. Once the array is in the air, it is also at the mercy of curious birds and squirrels. The varnish coat and strapping tape help preserve the helix tapes.

There is a mounting problem for the 32-ft long 40-m, 1/4-wave matching section. The 450-ohm open wire runs from the helix feed point, up to the top of the vertical spreader and back down to the tri-gamma match coaxial cable feed point. TV type screw-eye insulators along the vertical spreader hold the ladder line in place.

Boom support in Phase II was by suspension from a T bar at the mast. The Phase III boom uses a truss system support.

Assembly, Tuning and Testing

Each of the three elements measure 23 feet across the spreaders and 18 feet for each side of a quad loop. A critical series of holes drilled in the

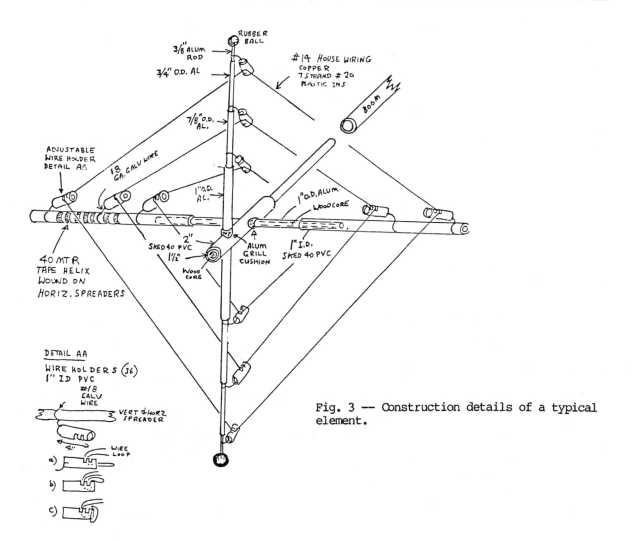

Fig. 3 — Construction details of a typical element.

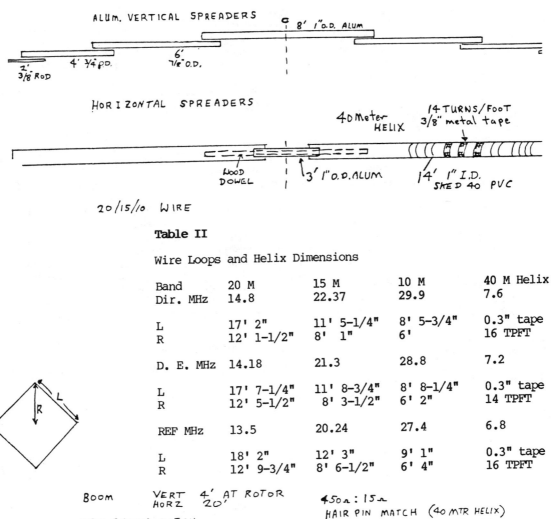

ALUM. VERTICAL SPREADERS

8' 1" O.D. ALUM

2' 3/8 ROD 4' 3/4 O.D. 6' 7/8" O.D.

HORIZONTAL SPREADERS

40 Meter HELIX   14 TURNS/FOOT 3/8" metal tape

WOOD DOWEL   3' 1" O.D. ALUM   14' 1" I.D. SHED 40 PVC

20/15/10 WIRE

Table II

Wire Loops and Helix Dimensions

Band	20 M	15 M	10 M	40 M Helix
Dir. MHz	14.8	22.37	29.9	7.6
L	17' 2"	11' 5-1/4"	8' 5-3/4"	0.3" tape
R	12' 1-1/2"	8' 1"	6'	16 TPFT
D. E. MHz	14.18	21.3	28.8	7.2
L	17' 7-1/4"	11' 8-3/4"	8' 8-1/4"	0.3" tape
R	12' 5-1/2"	8' 3-1/2"	6' 2"	14 TPFT
REF MHz	13.5	20.24	27.4	6.8
L	18' 2"	12' 3"	9' 1"	0.3" tape
R	12' 9-3/4"	8' 6-1/2"	6' 4"	16 TPFT

BOOM   VERT 4' AT ROTOR   HORZ 20'

TRI-GAMMA MATCH

	L	S	C (VAR)	(CAL)
20	35"	2"	100	80
15	27"	1.5"	50	30
10	18"	1"	75	60

COMP. CAP 250 μμfd VAR (150 FXD)

450Ω : 15Ω HAIR PIN MATCH (40 MTR HELIX)
2" STRAP TUNED AT 24"
28"

3/4"

Fig. 4 -- Dimensions for the vertical and horizontal spreaders.

boom assembly determine equal vertical alignment for the three elements. A large vice rigged with a carpenters square helps to align the hub holes for the spreaders, as well as the boom holes for key eyebolts. The boom eyebolt holes may be relocated as needed to improve alignment of the elements. An overall coat of spar varnish added some weight, but helped with weathering.

The four tunable sections of each of the three elements must be tuned individually before assembling the beam. Each element is tuned in a horizontal attitude, and mounted on temporary poles (clothesline) about 8 feet above ground (Fig. 9). Each wire loop is resonated using a dip meter. No stubs are used. See Table II for tuning frequencies. Dip-meter coupling to the helix is accomplished with a link of no. 16 enameled wire. Two turns folded over the helix, a short twisted pair link, and two turns for the coupling to the dip meter coil does the job. Wind each helix winding a foot longer than the Table III dimensions, then peel off turns to the desired frequency. The center element is mounted on

the mast first. Then the hollow boom sections are bolted in place. The director and reflector assemblies are added, using various helper support structures. A 10-foot "A" ladder is a must for this part of the job. After the array is assembled, but before erecting the tower, perform another series of dip meter checks to verify previous work.

This assembly proved too unwieldy to "gin pole" onto an erected tower. The three elements were assembled on the boom and mast with the tower folded over such that the rotor section was 15 feet above ground. Final tuning was performed with the array mounted on an 18-ft section of the tower with the bottom of the wire loops just clearing a nearby chicken coop roof.

An SWR bridge was inserted at the coaxial feed point of the tri-gamma match network. An extension of the SB-104A transceiver key circuit allowed keying and tuning at the antenna.

The tri-gamma match tuning scheme from Bill Orr's book worked perfectly. An SWR bridge is mounted in the coaxial feeder about three feet from

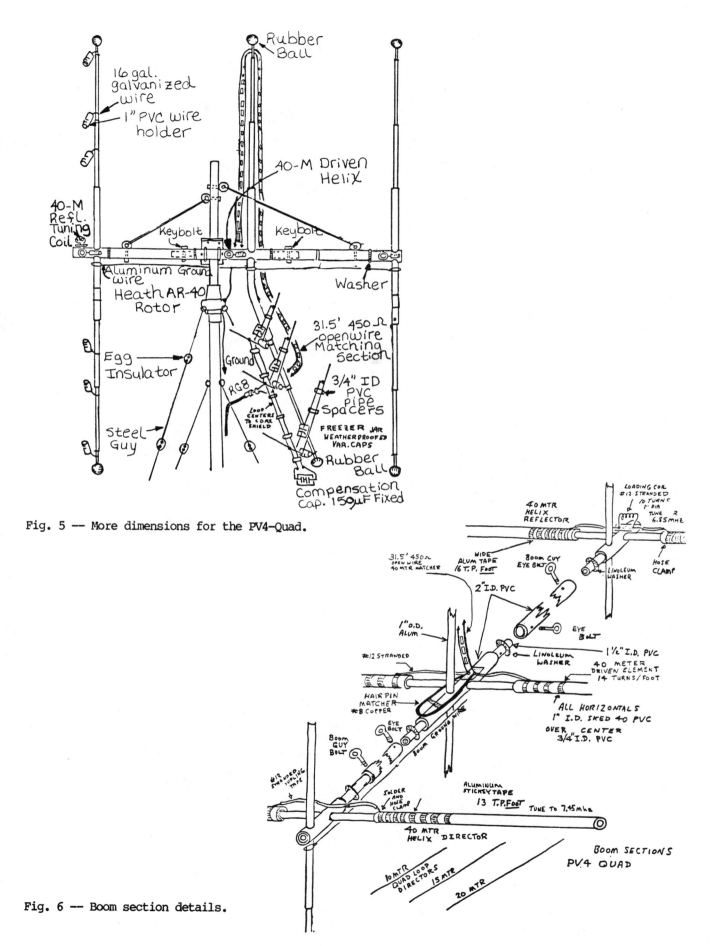

Fig. 5 — More dimensions for the PV4-Quad.

Fig. 6 — Boom section details.

Fig. 7 -- Typical hub. Note the loading coil tuned reflector helix.

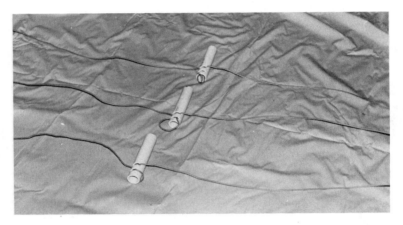

Fig. 8 -- The wire loops are firmly anchored by a three-step lock-in loop through the PVC holder.

Fig. 9 -- Each element is tuned in a horizontal attitude and mounted on temporary poles 8 feet above the ground.

the feed point at the 15-m driven element loop bottom. The sequence of tuning steps is as follows:

1. Set the variable 250-pF compensator capacitor to 200 pF.

2. Adjust the 10-m tap point distance from the bottom apex of the loop, and the variable capacitor (both for minimum SWR).

3. Do the same for the 15-m tap point and tuning capacitor.

4. Then do the same for the 20-m tap point and tuning capacitor.

5. Finally, the compensator capacitor is used to recheck each band for minimum SWR.

6. Then start over again with the 10-m adjustments.

Each of the variable capacitors is mounted in a plastic freezer jar and weatherproofed with duct tape and varnish as shown in Fig. 10.

Adjustments of the 40-m helix involved peeling off turns and resoldering turns at the outer ends. A tuning coil at the center of the reflector and director helix windings provided additional adjustment. Most of the 40-m tuning was accomplished with the individual elements mounted at 4 to 8 feet above ground in a horizontal attitude.

Performance Evaluation

The PV4-Quad was used for five years atop a 34-ft Heights aluminum tower at my Illinois location (Fig. 11). Performance data consists of hundreds of worldwide contacts on all four bands. Numerous QSOs involved tests using an HP 7035 XY plotter. The plotter X scale was driven from the offset voltage of the beam rotator controller circuit. The receiver "S" meter circuit fed the Y scale. See Fig. 12. Fig. 13 shows the X-Y plotter interface. The XY data was transposed to coordinates of the familiar cardioid pattern. Typical SWR measurements for a 20-, 15-, 10-, and 40-meter band are shown in Fig. 14. Compare this data to that shown in Fig. 15 for the PV4-Quad. Fig. 16 displays a radiation pattern for the PV4-Quad.

Fig. 10 — Freezer jars weatherproof the gamma capacitors.

Fig. 11 — PV4-Quad on the author's former 34-foot tower.

Fig. 12 — X-Y plotter test setup. Note the
S-meter magnifier.

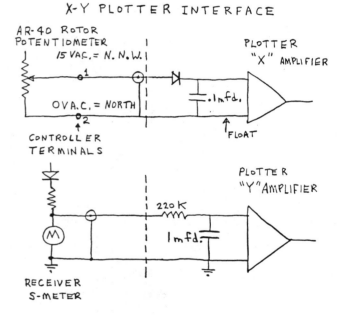

X-Y PLOTTER INTERFACE

Fig. 13 — The X-Y plotter interface.

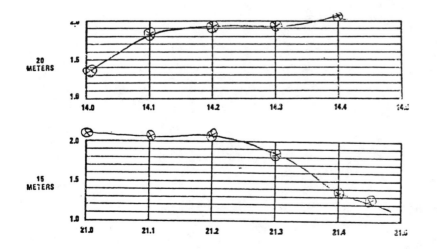

Fig. 14 — Typical SWR measurements for the
20-, 15-, 10-, and 40-meter bands.

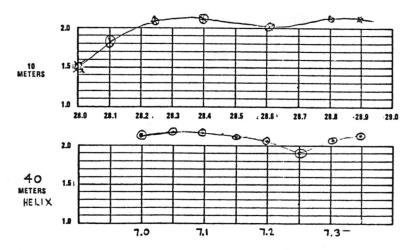

All bands 1:1 with tuner. PV4 Quad: One coax feed,
Measurement at 30 Watts CW mode, SWR bridge located
at Quad feed point.

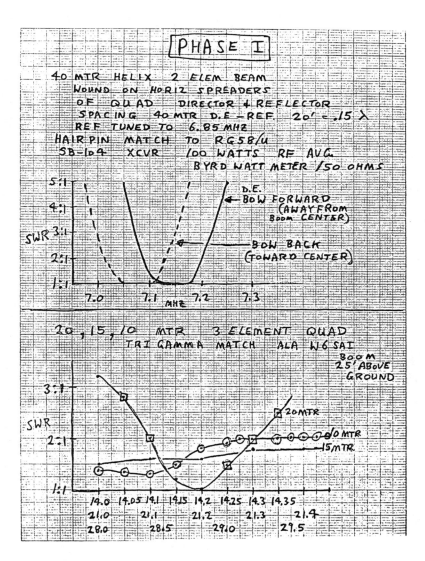

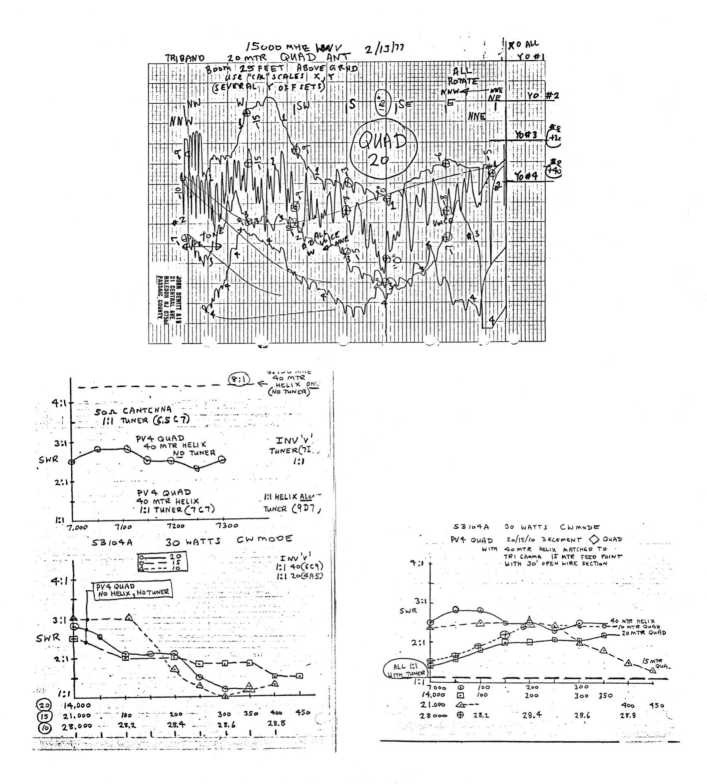

Fig. 15 -- SWR measurements for the PV4-Quad.

Summary

In the move from Illinois to New Jersey, the helix tapes were damaged. New tape windings are now being evaluated. The present mounting on a test tower beside a former chicken coop serves tuning and pruning very well. A nearby inverted V antenna is used for comparison measurements.

Note the overhanging oak trees in the Illinois photos. Once again the quad design low angle of radiation has proved to work well at "low altitude."

During the summer of 1984, the PV4-Quad was again reassembled atop a new tower. For many of the parts, it was the sixth time of disassembly and reassembly. The PVC sections have proved to be much more serviceable than I ever planned at the start of the project.

References

\1 Meyers, R., W1FBY, "The HW-40 Micro Beam," **QST,** Feb. 1974.

\2 Orr, Bill, W6SAI, **All About Cubical Quad Antennas.**

\3 Simon, B., W2UNN, "Surface Area of Quad Antennas," **CQ,** July 1967.

\4 Lindsay, J. E., WØHTH, "Quads and Yagis," **QST,** May 1969.

\5 Bergren, L., WØAIW, "The Multi-Element Quad," **QST,** May 1963.

\6 Schultz, John, K3EZ, "A New Look at Helically-Loaded Antennas," **CQ,** April 1976.

\7 Gooch, et. al., "The Hairpin Match," **QST,** April 1962.

\8 Grammer, George, W1DF, "Simplified Design of Impedance Matching Networks," **QST,** March 1957.

\9 Perrott, J. H., W4FRU, "Quad vs Triband Yagi," **QST,** February 1971.

Table III

Tape Helix Test Windings
0.3-inch wide copper tape on 1.25-inch OD PVC

Test #	Tape Length	Turns	Helix Length (Feet)	Frequency (MHz)
1	50'	1 TPinch	12.5	18.5
2	50'	0.8 TPinch	13	17.6
3	50'	16 TPfoot	9	19.5
4	50'	6 TPfoot	14.5	20.5
5	50'	6 TPfoot	20	15.5
6**	75'	16 TPfoot	14	14.0

**1/2 40-meter helix

E1	35	6 TPfoot	24	14
E2	35	15 TPfoot	6	28

Note: Copper foil tape with CONDUCTIVE pressure sensitive adhesive. Two suppliers are:

Prehler Electrical, Inc.
2300 N. Kilbourn
Chicago, ILL 60639
tel. (312) 384-6100
(3M Type 1181 Electrical Tape)

Chomerics, Inc.
77 Dragon Court
Woburn, MA 01801
tel. (617) 935-4850

List of Materials

Aluminum tubing: 6 and 8 ft lengths of 1-7/8 inch, and 3/4 inch OD.
PVC pipe (must be "Schedule 40" grade):
    6 20 foot lengths 1 inch ID
    3 12 foot lengths 2 inch ID
    4 12 foot lengths 1 1/2 inch ID
Wood clothes closet pole 1-3/8 inch dia., 12 foot length
500 feet no. 14 stranded copper house wiring (7 strands, no. 20 gauge)
33 feet of 450-ohm open-wire ladder line
6 feet of no. 8 copper rod (hairpin) RG8 coax, aluminum ground wire, steel guy wire
Miscellaneous bolts, clamps, and so on.
A Heath AR-40 rotor
Poly clothesline and pulleys for sky-hooks

Weight of Materials

500 feet of copper wire	5 lbs
20-foot boom, PVC and wood	25 lbs
3 26-foot aluminum verticals	19 lbs
3 PVC horizontals	26 lbs
Boom center vertical yoke	15 lbs
Coax and misc. clamps, boom guy	10 lbs
	100 lbs
Rotor	15 lbs
Steel tubing, center mast and T bar	10 lbs
	125 lbs

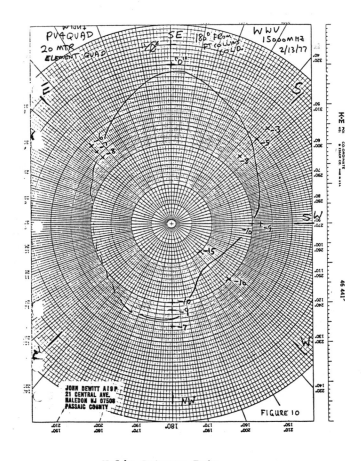

Fig. 16 — Radiation pattern for the PV4-Quad.

Helix Antenna Data

6 test windings on 20 foot 1.25-inch OD PVC pipe (Schedule 40). Copper tape was 0.3-inch wide (3M stickytape, 1-inch wide, cut into three pieces). Stock reel, 1" x 18 yards.

Two dip meters were used: General Radio (vacuum tube) and a Heath solid state.

	G. R.	Heath
1) 50 feet of tape, 1 turn per inch, 48" tape per foot, 12.5 foot long coil	18.6 MHz	18.5 MHz
2) 50 feet of tape, widened spacing 1/2 coil length, 6' coil, 1 turn per inch and 7' coil 0.8 turn per inch, both a 13' coil	17.6 MHz	17.7 MHz
3) 50 feet of tape, close spaced, 9' coil length, 16 turns per inch	19.5 MHz	20.0 MHz
4) 50 feet of tape (29' excess coiled), 14.5' coil length, 6 turns per inch, wide spaced	20.5 MHz	21.0 MHz
5) 50 feet of tape, 20' coil length, 6 turns per inch	15.5 MHz	15.5 MHz
6) 75 feet of tape, 14' coil length, 16-17 turns per foot	* 14.0 MHz	14.0 MHz *

Estimates: 35' tape (0.3" wide), 6 turns per inch, 24' long — 14 MHz
         35' tape (0.3" wide), 15 turns per foot, 6' long — 28 MHz

Cubical-Quad Antenna Design

By Norris G. Boucher,* W3GNR

*913 Chestnut Avenue, DuBois, PA 15801

Diamonds, deltas, Swiss and squares: These various types of quad antennas have provoked much speculation until direct gain comparisons between cubical quad and Yagi antennas settled crucial arguments in favor of the Yagi antenna.\1   Many prospective antenna builders were so disillusioned by the report that the cubical quad is seldom considered for directive antenna applications. This is unfortunate because the cubical quad can be an inexpensive alternative to the mono-band Yagi or specialty antennas, especially at frequencies that allow construction from locally available materials.

Unlike for the Yagi antenna, the selection of quad antenna designs is limited and often unreliable. Attempting to define satisfactory cubical quad antenna models for various spacings and numbers of elements, I developed the information presented in this article by experiments, which I will describe.

Project Goals

My first attempts to design and build quads of various types generated baffling problems. Frequency scaled cubical-quad antennas usually worked poorly, as did successive duplicate models at other sites. Taking reliable antenna gain measurements on location also proved to be discouragingly difficult.

The practical goal for this series of experiments was to produce a model set of low SWR, frequency scaleable quad antennas with three to eight elements and maximized gain. For simplification, only models with one reflector and equal-length directive parasitic elements were planned (several different models were tested). Other curious effects required explanations, and I hoped to discover their cause as my work progressed.

Element Materials

I have constructed dozens of antennas with little regard for the type of insulation, if any, on the wire. So it was interesting to learn that the resonant length of insulated quad elements is significantly different from their uninsulated counterparts. All information found in this article applies to **uninsulated** element quad antenna designs.

I have not measured the extent of this effect relative to resonant frequency, gain or quality of insulation (surface corrosion or thickness). Some trials indicated that insulated-wire elements lowered the operating frequency of the quad antenna perhaps by two percent. So note carefully that insulated-quad-element designs may not exhibit other effects reported here.

Element Diameter Scaling Experiment

The cubical-quad antenna has a relatively broad operating bandwidth. This is one of its better qualities. Unfortunately, that quality makes it difficult to determine optimum quad-element design k factors (ft - MHz). While trying to tablulate k factors for different elements from published designs, it became obvious that many experimenters, like myself, were floundering for lack of consistent design information.

Applying frequency scaling techniques added to the confusion. Researchers have found it necessary to consider the thickness of an element (expressed as an element diameter-to-wavelength ratio,* d/l, in consistent units) to successfully frequency scale model antenna designs.\2   Furthermore, the information I found on the quad-element d/l ratio effect was virtually a sidebar in some quad antenna literature.\3

An experiment to define the extent of the d/l effect on the circumference of cubical-quad elements was performed in this manner. Two open circular quad elements of equal length were placed in parallel, physically and electrically, and connected to a 50-ohm coaxial line. One loop was axially rotated 90 degrees from the other to roughly define a spherical cage. That configuration presented the proper load impedance to the feed line at resonance, decoupled the elements to minimize mutual impedance effects and increased the Q factor of the antenna.

* This concept is sometimes found in other references as a wavelength-to-diameter ratio, l/d. This is the inverse of d/l.

Increased Q was imperative since a single driven quad element exhibited a very shallow SWR curve at 52 +/- 2 MHz, the frequencies of measurement. When fed at the bottom or top of the loops, this quad loop cage makes an excellent, broadband, omnidirectional, horizontally polarized antenna.

Crossed-loop quad antennas were built and resonated, using no. 6 to 14 AWG bare copper wire. For each gauge wire, SWR measurements were made at numerous frequencies, and the values plotted to generate a characteristic SWR curve. From those curves, the frequency of the lowest SWR was selected, the loops measured and a k factor was calculated for each gauge element.

As a group, the resonant k factors (Kr) increase with an increase in element diameters, although the specific values are not sufficiently accurate for design purposes. The measurement error of my equipment exceeded the percentage differences of the k factors. It was confirmed that the k factor approached Kr = 1005, a value given for resonant quad elements in many references, as the d/l ratio decreases (thinner elements).\4

At least two adequately accurate resonant element k factor values resulted from further work in this series of experiments. It was found that eighth-inch bare aluminum wire has a Kr = 1020 at 144 MHz; 0.064 inch (no. 14 AWG) bare copper wire has a Kr = 1005 at 50.125 MHz. This, and information from another source, make it possible to construct a graph (Fig. 1).\5 This graph may be used as a guide, until more data is available.

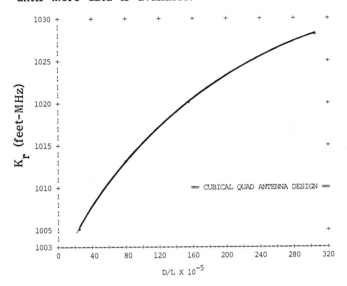

Fig. 1 -- Resonant element k factor vs. element diameter-to-wavelength ratio. This curve was generated from limited data to serve as a guide, until more accurate values are available. Note that the circumference of quad elements increases with increasing element diameter.

Element Diameter Scaling

To use Fig. 1, calculate the d/l ratio of the element diameter selected. For example, if you choose no. 14 AWG bare copper wire for use at 50.125 MHz, as I did:

$$\frac{d}{l} = \frac{\text{diameter (inches) x } f \text{ (MHz)}}{11803 \text{ (in-MHz)}} \quad \text{(Eq. 1)}$$

$$= \frac{0.064 \times 50.125}{11803}$$

$$\frac{d}{l} = 0.00027$$

Then, referring to the graph, locate the k factor for that d/l ratio. It should be Kr = 1005.

Using two k factors, a model correction factor (Ck) can be calculated. For example, Kr = 1020 (model) at 144 MHz and Kr =1005 (new) at 50.125 MHz, then calculate:

$$C = \frac{k \text{(new)}}{k \text{(model)}} \quad \text{(Eq. 2)}$$

$$= \frac{1005}{1020}$$

$$Ck = 0.985$$

To complete the scaling process, multiply the individual model element k factors by Ck, then divide by the selected operating frequency. The result is the length of the selected quad element.

Quad Antenna Model Experiment

Two computer-designed, five dBd, two-element Yagi reference antennas were deployed about 100 ft apart.\6 One was excited by a 144-MHz crystal controlled exciter. The other, used as a reference, was placed two wavelengths to one side of the antenna test area for the quad models. Each measurement was taken relative to the fixed reference antenna to compensate for system and signal path changes.

Test signals are routed through feed lines of equal length from the reference and test antennas, a two-port coaxial line switch, a precision attenuator or an SWR/power meter. This depends on the measurement being taken. Signals are finally fed to the 144-MHz receiver where a variable scale, slide-back voltmeter is attached to the receiver for measurements. I used an antenna tuner in the model antenna feed line during gain measurements.

The quad antenna test jig consisted of two vertically separated and parallel wooden strips. They were drilled at many places along their length to allow various trial element spacings. U-shaped portions of the quad elements, made from eighth-inch solid aluminim wire, were slipped into those locations. The other side of the elements, shorter U-shaped portions, were attached by slip couplings to close the loops. The driven element was fixed at a circumference determined from previous tests.

Accuracy of Measurements

Although all nearby structures were wooden, the area set aside for modeling the quad antennas was selected more for convenience than theoretic suitability. Therefore, in my opinion, the gain of these models cannot be compared to other antennas within 3.0 dB. The meter I used to make these relative gain measurements had enough resolution to guarantee measurements within the accuracy of my standard attenuator (estimated at +/- 0.25 dB). Relative gain measurements were routinely repeatable to within +/- 0.2 dB, averaging around +/- 0.1 dB.

Selecting a Design Model

After researching available sources, I decided to model three different element spacings of 0.130, 0.165 and 0.195 wavelength.\7 These values are grouped around spacings that are reported to produce maximum gain and low SWR. The results are tabulated in Tables no. 1 through 3, respectively (the element k factors must be corrected for other d/l ratios).

Being a controversial subject, quad antenna gain was the first quantity plotted from the experimental data. Looking at Fig. 2, you will see how the relative gain of my quad models varied with boom length for each of three spacings. Except for one point on the 0.195-wavelength element-spacing curve (attributed to a noisy attenuator switch), a smooth curve was easily drawn through the data points.

There is a design consideration for these spacings that tends to make the selection of the number of elements less arbitrary. An oscillation in the parasitic element k factors is apparent in the tabulated data for even and odd numbers of directors. To analyze this phenomenon in more detail, graphs of model director (Fig. 3) and model reflector (Fig. 4) k factors were constructed.

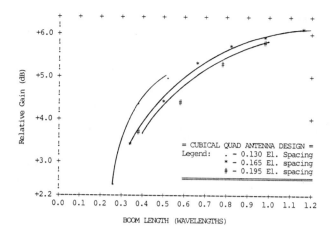

Fig. 2 — Model quad antenna relative gain vs. boom length. The slope of these collateral curves indicates the marginal gain of constant length director designs is approximately nine-tenths that of an optimized Yagi antenna for boom lengths in the region between 0.7 and 1.2 wavelengths. Optimized quad antennas should be expected to yield about the same marginal gain in that region.

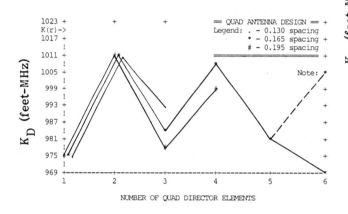

Fig. 3 — Director k factor vs. number of director elements. This graph clearly shows an oscillation in equal length director tuning for the first five directors. Notice the even numbered element K-factor groupings near the model resonant element k factor, Kr. Note: The 0.165-wavelength spaced model produced two gain peaks at the sixth director. The lower k factor produced more forward gain (+0.6 dB).

Clearly, these even and odd numbered element quad antenna models are quite different. Referring back to Fig. 3, notice that the four- and six-element quads have director k factors in the area of Kd = 1008, while three-, five- and seven-element designs have much lower k factors, grouped in the Kd = 984 region. Knowing the correct design k factors, you may reason that either odd or even numbered director quads could be built with equal ease. Theoretically that is true, but the resonant element k factor in these examples is Kr = 1020. An error in director length measurement of several tenths of one percent longer will noticeably decrease antenna gain at the design frequency.

This problem with an even number director element design is compounded by an accompanying oscillation in the quad reflector k factors. Looking at Fig. 4, notice that these designs are grouped closer

to a resonant element k factor. In this case, erring by making the element shorter would quickly reduce gain at the design frequency, but not as quickly as a lengthening error in the director elements.

This effect makes even numbered element quad antenna construction relatively critical. If you are willing to accept the increased accuracy requirements of this design, there is yet another consideration. When changing operating frequencies, the resonant frequency of the parasitic elements is more quickly approached.

Fig. 5 graphically displays the different element tuning for both even and odd numbered director element cases. Obviously, the gain and SWR curves of the even numbered director element quad will be sharper than the odd number designs. Given this data, the odd numbered director element quad antenna designs are a better choice.

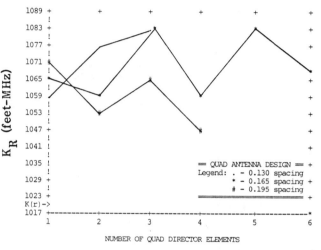

Fig. 4 — Reflector k factor vs. number of director elements. Although not as well defined, this graph shows a characteristic model k factor oscillation, as seen in the corresponding model quad director k factors (Fig. 3). Notice the even numbered element design k factors are grouped closer to a resonant element k factor, Kr.

Design Example

I used the information in this article to design and build five- and seven-element, 0.165-wavelength spaced, 50.125-MHz quad antennas. My choice of element material was no. 14 AWG solid copper wire stripped of insulation. Use Eq. 1 to compute the d/l ratio for this wire, d/l = 0.00027, as before. Then go to Fig. 1. Finding that d/l ratio on the horizontal axis, tracing that point upward to the curve, then left to the vertical axis locates Kr = 1005 for this size wire. Next, use Eq. 2 to compute the correction factor, Ck = 0.985, also as before.

Consulting Table 2 for five-element model k factor values, the reflector k factor (Kr) is 1083, and the k factor of the directors (Kd) is 984. Now,

$$\text{Element length} = \frac{C(k) \times k(model)}{f(MHz)} \text{ (ft)} \qquad \text{(Eq. 3)}$$

So,

$$\text{Directors} = \frac{0.985 \times 984}{50.125} \text{ (ft), or}$$

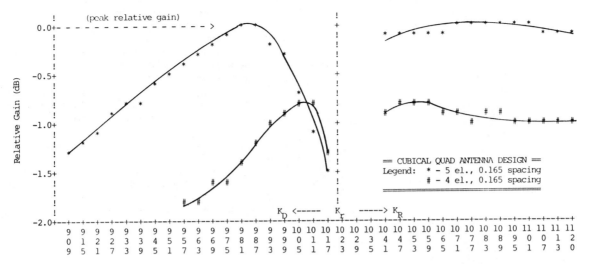

Fig. 5 — Relative gain vs. quad element
length. These curves demonstrate the difference
in lengths (for d/1 = 0.00027) of quad directors
and reflectors, which produce maximum gain for
even and odd numbered director model quad antenna
designs of 0.165-wavelength element spacing.

Directors = 19.34 ft (19' 4").

And,

Reflector = $\dfrac{0.985 \times 1083}{50.125}$ (ft), or

Reflector = 21.28 ft (21' 3-3/8").

Finally,

Driven element = $\dfrac{0.985 \times 1020}{50.125}$ (ft), or

Driven element = 20.04 ft (20' 1/2").

Now that the element lengths have been determined, the element spacing, boom length and
spreader length are calculated.   Thus,

Spaces = $\dfrac{Spacing(waves) \times 983.58}{f\ (MHz)}$ (ft),

$\qquad = \dfrac{0.165 \times 983.58}{50.125}$ (ft), or

Spaces = 3.24 ft (38-7/8").

Next,

Boom = (no. of elements – 1) x spaces (ft)

$\qquad$ = (5-1) x 3.24 (ft), or

Boom = 12.96 ft (12' 11-1/2").

And finally,

Spreader length = $\dfrac{Element\ length}{2.828}$ (ft),

For example, the driven-element spreader length

$\qquad = \dfrac{20.04}{2.828}$ (ft), or

Spreader length = 7.09 ft (7 ft 1-1/32 in) between
wire holes for the driven element.

Calculate the reflector and director spreader
lengths in the same manner.

An interesting feature of this model is that it
can be modified to make a five- or seven-element
convertible quad antenna. If you look at Table 2
again, notice the relatively minor difference between the director k factors of the five- and
seven-element model quads. Using either k factor
will produce a satisfactory antenna. So this design
can be changed by adding two director elements to
accommodate changing circumstances, like portable
and fixed operation with attendant differences in
gain.

Sample Construction Materials

The spreaders used to build the 50.1-MHz quad
test antenna were lengths of 1/2 x 3/4 inch, rectangular, wooden doorstop molding. Each was gener-
ously coated with several layers of high grade, oil
base (not latex) paint. This wooden material is
inexpensive and available in 16-ft lengths. If you
wish to use it for spreaders greater than eight but
less than twelve ft, cut the strip in half. Overlap
the halves to the proper length and glue the pieces
together to increase their strength.

The spreaders were individually held to the
boom with U bolts which, when disassembled, allow
the elements to be folded for transportation. Also,
the spreaders were aligned vertically and horizontally to produce diamond-shaped elements. With
the feed point for horizontal polarization occurring
at the end of the lower driven element spreader, the

feed line was conveniently dressed along that arm, back the boom and down the mast. The boom material was sections of steel TV mast, and a metal, square electrical box cover was drilled and fitted with U bolts to hold the boom to the mast. The mast was a 10-ft length of 1-1/4 inch thinwall EMT conduit. So constructed, the antenna was light in weight, and two of these quad antennas were easily turned by a TV antenna rotator.

The Obligatory Comparison

Experimenters find it difficult to justify a number of persistently reported quad-antenna characteristics. To explain those, I believe it is necessary to look beyond the physical quad antenna and to integrate fact, theory, observation, circumstance and environment. Unfortunately, that is beyond the scope of this article. However, I have no argument with the claim that the horizontally polarized Yagi antenna is superior to a quad antenna for competitive communications.

There is one quad antenna characteristic I would like to discuss. Consider that the closed quad loop has no physically constrained or inherent polarization. It follows, then, that an array of quad loop parasites will have equal gain with any angle linear polarization. Therefore, a quad antenna is dependent on a single element, the driven element, to fix its majority polarization. In comparison to the Yagi, the quad antenna lacks the cascaded polarization selectivity of Yagi parasites and does not competitively reject off-horizontal wave components.

These facts are experimentally observable as an anti-fade characteristic of the quad antenna. Over a period of months, I observed reduced signal fading on a 150-mile VHF communications path from a station using a high-gain quad antenna array. Numerous stations using Yagi antennas were available for comparison during the same time period. Then for further verification, the local Yagi was replaced by a quad antenna. Signal fading from other stations was reduced. All other things being equal, it is reasonable to assume that the observed signal fading was caused, in part, by polarization rotation and that the anti-fade characteristic of the quad antenna was because of its parasitic element response to nonhorizontally polarized wave components.

This discussion and earlier remarks define areas of communication where the quad antenna may be expected to perform well. For casual or low-budget communications, construction of a quad antenna yields economy of resources, reduced fade characteristic, increased compatibility with other antenna types, respectable gain and broader operating bandwidth than the Yagi antenna.

The quad antenna may even have its niche in competition with single polarization Yagi antennas. Since quad antenna polarization is dependent on the point where the driven element is excited, moving the feed point in various ways allows the user to take advantage of the prevailing optimum polarization with no loss in antenna gain. This chameleon-like quality, on balance, may make the quad antenna very competitive in some circumstances.

Conclusion

The practical goals of these experiments were met. Using the information presented in this article, a quad antenna that required no tuning was designed and built for 50.125 MHz. In use, the pattern appears clean, the SWR is excellent and the gain appears to be in line with the model. By taking care to duplicate the models, using bare elements and nonmetallic spreaders and correcting for differing element d/l ratios, good results can be expected.

The narrowest element spacing may be of greater interest for quad antenna designs in the upper end of the HF spectrum. Translating the boom length of the longest 0.130 wavelength spaced model into feet for example, a 17.8-ft boom is necessary to accommodate five elements at 28.7 MHz. For economy of elements, rather than boom length, the wider element spacings are attractive at VHF. There for example, a 19.4-ft boom will hold seven either 0.165- or six 0.195-wavelength spaced elements for 50.1-MHz quad designs.

Do not be concerned with SWR, unless it differs greatly from that of the model at your design frequency. SWR is not an indication of gain, nor an indication of the frequency where maximum gain occurs. No attempt was made to match the driven element to the feed line, when modeling, so the listed values of SWR are intrinsic to these specific models.

These quad antenna models were gain maximized for equal length quad directors. Front-to-back (f/b) ratio was never a factor in these experiments. While it may be impressive and sometimes desirable to observe large f/b ratios in use, the factor is not a valid indicator of forward gain. Indeed, some forward gain must be sacrificed to optimize the f/b ratio.

Several dual reflector quad antenna models were modeled, but the gain improvements were not thoroughly investigated (see notes in Tables 1 through 3). However, a second quad reflector looks promising instead of a fourth quad director for increased gain. There it also preserves the desirable odd numbered director element design format.

References

\1 W. Overbeck, N6NB, "Quad vs Yagi Revisited," **Ham Radio**, May 1979, p. 12.

\2 P. Viezbicke, "Yagi Antenna Design," **NBS** Technical Note 688, U. S. Department of Commerce, Washington, DC, December 1976.

\3 **The ARRL Antenna Handbook**, 13th edition, Chapter 4, "The Quad Antenna," p. 159, second paragraph.

\4 See Ref. 3.

\5 See Ref. 3.

\6 J. L. Lawson, W2PV, A Series: "Yagi Antenna Design," **Ham Radio**, 1980.

\7 William I. Orr, W6SAI, **All About Cubical Quad Antennas**, 2nd edition, 1970.

Table 1.
quad model: 0.130 wavelength-element spacing.
Number of elements (including driven), element
design k factors (ref: Kr = 1020 ft-MHz), boom
length (wavelengths), gain (relative to a two-
element Yagi), and SWR (:1).

Number of Els	quad Model k factors Reflector	All Dir's	BoomL	G(dB)	SWR
3	1059	975	0.260	2.3	1.3
4	1077	1011	0.390	4.4	1.1
4*	1101	981	0.390	3.0	1.3
5	1083	993	0.520	5.0	1.3

*Note: Two reflectors, one driven, plus
directors.

Table 2.
quad models: 0.165 wavelength-element spacing.
Number of elements (including driven), element
design k factors (ref: Kr = 1020 ft-MHz), boom
length (wavelengths), gain (relative to a two
element Yagi), and SWR (:1).

Number of Els	quad Model k factors Reflector	All Dir's	BoomL	G(dB)	SWR
3	1065	975	0.330	3.5	1.4
4	1059	1011	0.495	4.4	1.2
4*	1065	987	0.495	3.7	1.3
5	1083	984	0.660	5.2	1.2
6	1059	1008	0.825	5.7	1.1
7	1083	981	0.990	5.9	1.3
8	1068	969	1.155	6.1	1.1
8#	1068	1005	1.155	5.5	-
8*	1053	981	1.155	6.3	1.3

Note: #As above, directors at another gain peak.
       *Two reflectors, one driven, plus
directors.

Table 3.
quad models: 0.195 wavelength-element spacing.
Number of elements (including driven), element
design k factors (ref: Kr = 1020 ft-MHz), boom
length (wavelengths), gain (relative to a two-
element Yagi), and SWR (:1).

Number of Els	quad Model k factors Reflector	All Dir's	BoomL	G(dB)	SWR
3	1071	975	0.390	3.7	1.2
4	1053	1011	0.585	4.4	-
4*	1059	987	0.585	3.6	1.4
5	1065	978	0.780	5.3	1.2
6	1047	999	0.975	5.8	1.4
7*	1059	999	1.170	6.2	1.1

*Note: Two reflectors, one driven, plus
directors.

Log Periodic Arrays

A Wide-Band, Low-Z Antenna — New Thoughts on Small Antennas

By Ken Heitner,* WB4AKK

*2410 Garnett Court, Vienna, VA 22180

Today's radio amateur has the opportunity to select flexible equipment that will operate on a number of amateur frequencies. Confronted by the dilemma of living in crowded urban areas, these same amateurs don't have the acreage to install separate antennas for each band. This clearly creates a need for a small, versatile antenna -- one that is physically compact, but having the bandwidth to cover several amateur bands. If such antennas provided directivity and were inexpensive, they would also be attractive. Most amateur thinking on antennas, however, is along "traditional" lines. Antennas are typically quarter- or half-wavelength resonant structures, and matched directly to 50- or 72-ohm feed lines.

A few amateurs have thoughtfully worked on solving the problems of small antennas, indicating that their size can be reduced. The developed designs, however, are for resonant antennas with limited bandwidth.\1

Characteristics of Small Antennas

With a reduction in size, the antenna radiation resistance decreases and the reactance increases. Fig. 1 shows these radiation resistance changes for small dipoles and loops.

Lower radiation resistance presents two problems. First, the lower resistance must be coupled to the nominal 50-ohm transmission line. Second, as the radiation resistance becomes very low, resistive losses in the antenna become significant and antenna efficiency goes down. Resonant antennas have previously been built with radiation resistances in the order of 3 ohms with good performance, and mobile whip antennas for 80 and 160 meters routinely use lower radiation resistance levels. These antennas, however, operate with significant resistive losses.

When a dipole is less than its resonant length, its impedance has an increasing capacitive component (see Fig. 2). To resonante a short dipole, inductive loading must be added. This introduces additional resistive losses.

Small loop antennas exhibit an inductive impedance characteristic, but can be made resonant with a parallel capacitor. These antennas are desirable for this size category because large losses do not occur in the tuning capacitor.

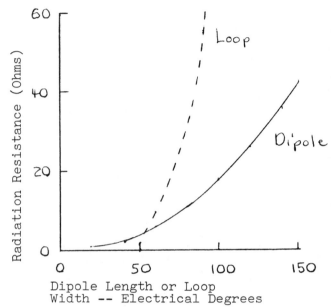

Fig. 1 -- Radiation resistance for small dipoles and loops.

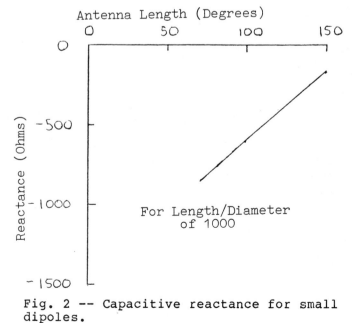

Fig. 2 -- Capacitive reactance for small dipoles.

Broadband Antenna Concepts

Various broadband antenna types have been developed with designs based on the fact that antenna structures defined by angles have inherently broadband characteristics. Examples include the discone and conical monipole.\2 Other designs are based on the concept of a repeating geometrically similar structure. The log periodic dipole array (LPDA) falls into this group. It consists of a series of dipoles that are geometrically related and connected to a common feed line. Elements resonant at or near the operating frequency actively radiate the signal. For a small broadband antenna design, the LPDA offers a potential approach. Antenna size can be reduced by modifying the individual elements. If the geometrical similarity and the element's electrical properties are preserved, the achievement of creating a smaller antenna can become reality.

LPDA Design

A description of conventional LPDA design is provided in **The ARRL Antenna Book** (1974 edition, pages 160 to 164). This design consists of tubular elements at right angles to the boom. The resulting antennas are large and heavy, requiring large rotators and sturdy towers for support. The need for a lighter, more compact LPDA led to an alternative design shown in Fig. 3.\3

The design approach shown in Fig. 3 is notable for several reasons. The dipole elements are raked forward, reducing the turning radius of the array, and the elements are wires held in tension by the array structure. Because the array structure is a boom and a single cross piece, it is lighter and has a low cross-sectional area. Also, the number of wire elements can be easily increased if desired. Since the array geometry for the LPDA departs from the published data in **The ARRL Antenna Book**, I cannot evaluate it's performance.

Thoughts on Compact LPDA Designs

Several considerations come to mind when trying to evolve a compact LPDA for amateur use. Although the basic dipole element can be physically shortened, it requires retuning to some new frequency. On the other hand, if the dipole element was reduced in size (by folding it into a half-wave loop), no retuning would be necessary. With the loop element having directivity, it could offset gains loss by the use of close element spacing.

Low radiation resistance of the half-wavelength loop can be raised by folding the loop element. This procedure is analagous to a folded dipole. The nominal impedance of the half-wave loop is about 10 ohms and it would be raised to about 40 ohms with a second folding. The second folding could be accomplished by constructing the wire elements out of 300-ohm twin lead in the manner of the classic folded dipole.

If the elements are constructed as a series of triangular loops, an array would appear as shown in Fig. 4. Although the loop elements overlap, the individual wires are insulated. There is no contact between elements at the crossover points. They are connected at their feed points by a parallel line feeder that has alternating "phasing" to each element set, and a balun would be used to assure a balanced feed impedance with respect to ground.

The triangular element shape also appears reasonable from a mechanical and structural viewpoint. The beam can be constructed with cross pieces for the smallest and largest elements. The remaining elements can be constructed of wires in tension, as is the design shown in Fig. 3.

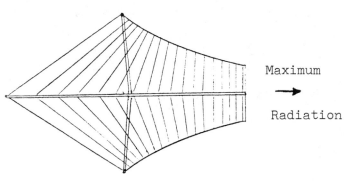

Fig. 3 -- Granger LPDA design.

Conclusion

It is possible to conceive of an antenna structure that possesses both wide bandwidth and compact qualities. In addition, it has directivity, lends itself to amateur construction techniques and though it has inherently low radiation resistance, the impedance level at the feed point can be transformed to the 50-ohm level.

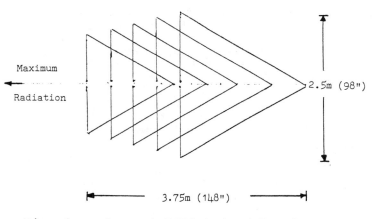

Fig. 4 -- Compact LPDA-triangular loops (20-30 MHz).

References

\1 **The ARRL Antenna Anthology**, 1978, pp. 22-29.
\2 **The ARRL Antenna Book**, 1974, pp. 298 and 299.
\3 Solutions for Telecommunications, Granger Associates, Santa Clara, CA (Recent).

Development of the W8JF Waveram: A Planar Log-Periodic Quad Array

By Jim Fisher,* W8JF

*Director, Compensation & Benefits, Air Products & Chemicals, Inc., P. O. Box 538, Allentown, PA 18105

Learn to use log-periodic principles to develop broadband antennas for the new post-WARC age, and discover some practical log-periodic quad arrays. This array won third place as a five-band design in the ARRL Antenna Competition (1984).

Radio amateurs have made extensive use of log-periodic dipole arrays (LPDAs), including the hybrid log-periodic Yagi (LPY) in which parasitic Yagi elements are excited by a small LPDA. However, most amateurs are not familiar with log-periodic approaches that differ from LPDAs. This article describes the development of one such alternative approach — log-periodic quad arrays (LPQAs), culminating in the W8JF Waveram. Several antenna types useful in Amateur Radio are also presented and different ways to use log-periodic principles in non-dipole applications are illustrated. With the advent of the new WARC bands, interest in the LPA's broadband capabilities should grow.

Characteristics of a Basic LP

Fig. 1 shows a conventional log-periodic array. It is sometimes seen in the form of a TV antenna. Note several characteristics of the LPA that are used in later discussions:

* The array is balanced; elements on one bay are offset by equally long opposing elements on the other.

* The two bays are fed 180 degrees out of phase through the booms, which are single-conductor feed lines.

* To provide uniform gain over a wide frequency range, adjacent elements and the spacing between them are scaled down progressively by a constant multiplier such as 0.9. Frequency coverage in the fundamental mode is from the low frequency (the longest element is a quarter wave), to the high frequency (the shortest element is a quarter wave).

* By virtue of the phase reversals and increasing element lengths along the feed line moving away from the feed point, the array has gain along its central plane (a horizontal plane as in the case of Fig. 1).

* The array has a front-to-back ratio in the direction of the smaller elements because the two bays "aim forward," as opposed to being in the same plane.

Deriving the LPDA from the Basic LP

Let's start to play with these concepts. If the two bays are moved together, the opposing elements look like dipoles. The booms may either be very close or replaced with a two-wire balanced transmission line twisted between each set of elements.

The result, shown in Fig. 2, is an LPDA. It has gain in the direction of the smaller elements, giving it front-to-back ratio.

Deriving the LPQA from the LPDA — a Misdirection?

A full-wave loop has a gain of several dB over a dipole. Why not replace each dipole with a loop (quad element) resonant at the same frequency?

I built such an antenna (represented in Fig. 3) for 6 to 10 MHz in the mid-1970s and described it in an article appearing in **CQ Magazine**.\1 The same con-

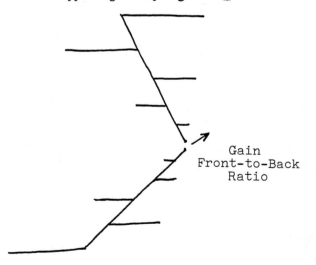

Gain
Front-to-Back
Ratio

Fig. 1 -- The conventional log-periodic array. The array is fed at the apex with balanced line, and exhibits gain and front-to-back ratio in the direction of the apex.

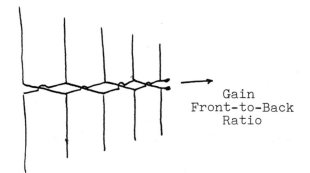

Fig. 2 -- The log-periodic dipole array (LPDA).

cept was described by W4AEO in **QST**.\2

As I studied the matter further, I began to realize that such an antenna, while performing well over a limited frequency range, has deficiencies that can be identified either through quad theory or log-periodic theory. Starting with quad theory, consider a full-wave quad loop as two half-waves in phase. We identify the gain as being derived from current maxima separated by a stacking distance. Normally the loop is fed only at one point, but the two current maxima are positioned by the half-wave distance from the feed point around the loop to the opposite point.

What happens to the LPQA constructed as shown in Fig. 3? At most, one loop is exactly a full wave; the others are progressively larger or smaller. The resulting current maxima in the other loops are not distributed in a symmetrical manner about a single central plane.

Even viewed from a distant point, the current patterns are not contributing efficiently toward a single pattern of forward gain. Furthermore, I found that the resistance and reactance patterns at the feed point (listed in my article) showed significant excursions even within the 6- to 10-MHz range. There are also potential problems, hinted at in my article, from nonfundamental resonances that can occur within the fundamental frequency range.

Try Again

Let's now reconsider the problem in light of log-periodic theory. Rather than thinking of an LPQA as an expanded LPDA, let us think of it as being derived from a log-periodic array resembling Fig. 1, but having four bays instead of two. The top two bays are collapsed into one LPDA and the bottom two bays are rearranged to form an identical LPDA. The four booms become not one, but two balanced transmission lines. The ends of the LPDA elements are folded to touch each other, and voila! An LPQA as shown in Fig. 4, with symmetrical current distribution and a concerted forward pattern, is formed.

My 6- to 10-MHz LPQA was converted to the improved version simply by cutting the top of each loop and inserting a second transmission line. Checks of resistance and reactance patterns showed significant improvements, with excellent SWR over a wide frequency range, while feeding coaxial cable with a 4:1 balun.

I thought the idea was original, but discovered that Collins had already obtained a U.S. Patent number on it (no. 3,279,159). I constructed several such antennas for SWL and Amateur Radio pur-

poses and described them in the 1978 **World Radio TV Handbook** (WRTH).\3

Dimensions given in Table 1 are for two antennas that are most likely to be of interest to Amateur Radio operators.

* A 6- to 18-MHz fixed LPQA wire antenna — effectively, a unidirectional 40-, 30-, 20- and 17-meter antenna fed with a coaxial cable through a 4:1 balun.

* A 9- to 18-MHz rotatable four-element LPQA — think of it as a 30-, 20- and 17-meter tribander based on standard ham radio quad hardware.

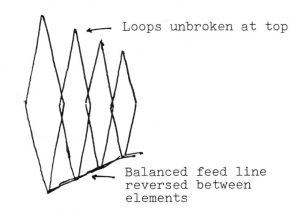

Loops unbroken at top

Balanced feed line reversed between elements

Fig. 3 -- First attempt at a log-periodic quad array (LPQA). Don't build it this way!

The fixed array can be constructed between two tall supports. If additional supports are available, two or more arrays can be constructed to cover various directions. Continuing the array to progressively smaller elements and shorter spacing or, at the bigger end, to larger elements and longer spacing, could broaden the already creditable 3:1 frequency range. At one time, I was using two of these arrays aimed about 60 degrees apart. Switching

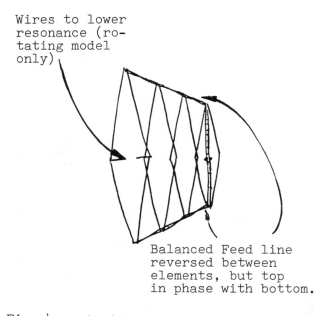

Wires to lower resonance (rotating model only)

Balanced Feed line reversed between elements, but top in phase with bottom.

Fig. 4 -- An improved LPQA feeds each loop in phase at its top and bottom.

from one to another with a coaxial-cable switch yielded a satisfying instant demonstration of pattern lobes. This antenna is effective and inexpensive to build, and could be used to cover 5:1 or greater frequency ranges anywhere from 160 meters on. The W4AEO article and my two articles contain some practical suggestions for methods and materials. With proper spacing, the array may function well on odd multiples of its fundamental resonances, yielding extremely broad continuous frequency coverage.

The rotatable array offered fair gain and front-to-back ratio, which is not easy to realize in a compact antenna over that continuous a frequency range. It was taken down shortly thereafter to make room for further experiments, some of which are described below.

Derivation of the W8JF Waveram

Let's derive another antenna two different ways to see if our results are consistent. As an objective, let's seek LPQA-type performance with a minimum boom length.

First, it occurred to me to reduce the boom length of an LPQA to zero, creating a planar log-periodic quad array which I call a W8JF Waveram (see Fig. 5). This is a broadband bidirectional array. Several versions have been tested at VHF, all displaying gain over a reference dipole. The first full-scale version, built while I lived in Michigan, was used to make two-way Amateur Radio contacts with hams on six continents while using an output power of two watts on 20 and 15 meters. For this article, I constructed and tested a ten-loop version (dimensions in Table II) in VHF scale model and full-sized HF versions. I used it with five watts output to make intercontinental QSOs on 10, 15, 20, and 30 meters. (Thirty-meter operation is below the fundamental frequency and is in another mode. This is not uncommon with LPs.)

Without a balun, SWR was no more than 2:1 for the 14-, 18-, 21-, 24-, and 28-MHz bands with a small capacitance across the feed points. Development for lower SWR should be possible, measuring SWR or R and X while varying spacing, element thickness, and transmission-line characteristics, and possibly using a tapped toroidal RF transformer. It should be noted that the SWR will change if the W8JF Waveram is used as part of a larger antenna array. To understand how this works and to confirm the approach, we will review the original LP model and try to develop a similar concept.

First, the bays in Fig. 1 are straightened into a single plane. The front-to-back ratio drops off to zero; the array becomes bidirectional, but continues to have gain in the plane perpendicular to the array. Sweep the tips of the elements to the center line and we have Fig. 6, with offset elements not meeting as loops. The feed lines are still single conductors. This is a regular LP, one of the traditional possibilities, although the sweep is more severe than we would normally see. (One with no sweep is called 90 degrees from the transmission line. A mild sweep of 60 degrees from the transmission line is more common in LPs than this 45-degree sweep.) An advantage of this design is that it can be constructed with standard 20-meter quad hardware; 13-foot spreaders provide a fundamental resonance just below 20 meters. I constructed VHF scale models and a full-scale HF version. Gain of

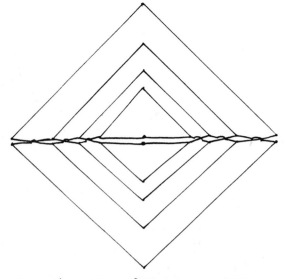

Fig. 5 -- The W8JF Waveram LPQA.

the VHF scale model was on the order of that of the equivalent fully balanced version. Reception was good and some DX was worked on the HF version, but the SWR was not as low as with the balanced transmission line. Again, there are ways to influence the SWR, and the Fig. 6 version may be better for some purposes.

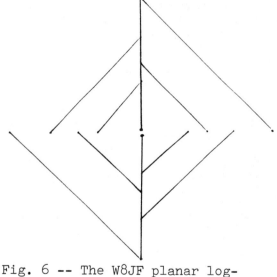

Fig. 6 -- The W8JF planar log-periodic array.

Try Again

To derive the fully balanced W8JF Waveram, we have to start with a four-bay LP. For instance, the bays should be spaced 90 degrees apart in an "X." We move the top bays together and the bottom bays together and we have opposing LPDAs. Sweep the element tips together to form continuous loops and the fully balanced version is formed.

Summary and Further Challenges

We have used log-periodic principles to derive several different practical broadband arrays for amateur use. All could use some further experimentation to optimize gain and SWR through element shape and sweep, scaling factors, shorting loops at the end of the transmission lines, and so on. Elements could be center loaded or linearly loaded with inductance or capacitance to lower or raise their resonant frequencies. All appear suitable for single coaxial feed through a balancing device, though some development work is needed before typical solid-state finals will run flat out into the load.

At one extreme (Fig. 4), there are fixed arrays that could cover wide expanses of the MF and HF spectrum, and are inexpensive to build provided two high supports are available (one support can be lower than the one supporting the biggest elements). The array in Fig. 4 varies from huge to fairly compact, depending on spacing and frequency coverage. A rotatable version of the Fig. 4 array has demonstrated gain and front-to-back ratio down to 9 MHz.

The W8JF Waveram is usable by itself as a very compact antenna. It offers gain over a dipole and broadband coverage with a single feed line. I would certainly prefer it to trap dipoles for most applications. Different versions have been tried, with help from my family. For example, the scaling factor has been varied from 0.7 to 0.8366, the array has been configured in a rectangle with the transmission line running parallel to the longer side, and a version has been built with the elements swept even more than the 45-degree model. Each version displayed good performance, to the extent that I would recommend the builder choose what is mechanically convenient. For example, the highly swept, flat-diamond model, mounted with the transmission line horizontal, yields a lower quad that is easier to handle when raising and lowering than the 26-foot diagonal model -- but it is vertically polarized! (The polarization of a Fig. 1, Fig. 5, or Fig. 6 array is perpendicular to the run of the transmission line; the polarization of a Fig. 2 LPDA is in the plane of the elements; the polarization of the Fig. 3 (don't build it!) or Fig. 4 arrays is horizontal with the transmission line(s) at the top and bottom and vertical with them at the sides.)

One purpose of building the W8JF Waveram was to create a compact, broadband building block and investigate whether it could be a basis for some more complex arrays. Actually, I first thought of it as a substitution for the forward element of a Fig. 4 array. Time has not permitted much work on these possibilities, but initial results indicate that the cookbook dimensions of parasitic dipole and quad elements may not be optimum for use with a W8JF Waveram as a driver, possibly because of the E and H-plane patterns of the array in the near field. I have also acquired a second parallel W8JF Waveram to show some activity as a parasitic element, with and without a tunable reactance across the feed point. In addition, I have obtained some hint of activity as a broadband sink for absorbing waves from the back as a second reflector and improving front-to-back ratio over a broad frequency range. The latter was attempted with a fixed resistance of 200 ohms, which may be too high. A more elegant solution would be attenuating open-wire line such as is used in non-resonant rhombics. Broadband reflectors such as corner reflectors or parabolic dishes have not yet been tried.

For those who have decided that these quad versions of the LP are not for them, I hope this has been an interesting exercise in LP approaches. It may be of assistance should you decide to devise your own arrays. If you experiment with my ideas, I would appreciate hearing of your results.

References

\1 Fisher, J., "Log Periodic Quad Array," **CQ**, February 1977.

\2 Smith, G., "Quad Log-Periodic Fixed Beam Antennas," **QST**, April 1977.

\3 Fisher, J., "Build Your Own Log-Periodic Quad Antenna," **World Radio TV Handbook**, 1978 edition.

Table 1
Dimensions for 6 to 18 MHz
Fixed LPQA

Element	Length of One Side of Loop Feet (Meters)	Spacing Feet (Meters)
1	83'9" (25.54)	10'0" (3.05)
2	71'2" (21.71)	8'6" (2.59)
3	60'6" (18.46)	7'3" (2.20)
4	51'5" (15.69)	6'2" (1.87)
5	43'9" (13.33)	5'3" (1.59)
6	37'2" (11.33)	4'5" (1.35)
7	31'7" (9.63)	3'9" (1.15)
8	26'11" (8.19)	

Note: The above lengths are for one-half of the loop (from transmission line to transmission line).

Dimensions for Rotatable
9 to 18-MHz Array

Element	Length of One Side of Loop Feet (Meters)	Spacing Feet (Meters)
1	36'9" (11.21)	9'9" (2.97)
2	36'9" (11.21)	7'10" (2.39)
3	34'8" (10.56)	6'4" (1.93)
4	27'11" (8.52)	

Note: To reduce the resonant frequency of the lowest loop, 16'7" (5.05 m) of wire is connected to each of the two outside corners, run along the horizontal spreader and doubled back on itself. Similarly, for element 2, 6'2" (1.89 m) of wire is connected to each of the outside corners and run along the horizontal spreaders.

Table 2
Dimensions for Planar Arrays

Distance along spreader from center for each connection

Element	Distance to Center Feet (Meters)
1	12'9" (3.88)
2	10'8" (3.25)
3	8'11" (2.72)
4	7'6" (2.27)
5	6'3" (1.90)
6	5'3" (1.59)
7	4'5" (1.33)
8	3'8" (1.11)
9	3'1" (0.93)
10	2'7" (0.78)

Note: In the fully balanced W8JF Waveram, wires are connected on both sides of each transmission line at each of the above points. In the W8JF Planar Log Periodic (PLP), wires are connected to alternating sides of each point as shown in Fig. 6. However, to balance the stress of the wires, it is suggested that a nylon cord be used for the runs without wires.

A Second-Generation Spiderweb Antenna

Dick A. Mack,* W6PGL

*600 Lockewood Lane, Santa Cruz, CA 95066

The original description of the Telerana (Spanish for Spiderweb) log-periodic antenna appeared the same time my son-in-law, Jim Nelson, N6EWP/VS6JA, departed as a career missionary to Hong Kong. Among our plans were skeds for frequent QSOs across the Pacific. Depending on the propagation, we expected to use 10, 15 and 20 meters. The Telerana appeared to be an almost ideal antenna for this service. It was light weight, rotatable, and presented low resistance to the wind. Electrically, log-periodic arrays have good forward gain and acceptable front-to-back ratio characteristics. The idea of switching from 10 to 15 and then to 20 meters without an antenna tuner was most intriguing. Also, the future 18- and 25-MHz bands could be covered without additional elements.

Now, three years later, I am convinced that the log-periodic antenna can rightly take its place among other high-performance beam antennas. Over this period, Jim and I have had numerous SSB QSOs, usually several times a week, with good results. A second array was built for my friend K7JIO two years ago.

The construction described here is based on the length and spacing of the antenna radiators described by Ansyl Eckols, YV5DLT, but with a different supporting structure.\1 Several improvements were made while building the second unit and are included here. The beam uses 13 wire elements with a 90% taper and a 0.05% wavelength spacing. As in all log-periodic dipole arrays, each pair of radiators is connected to the feed line in a transposed fashion with respect to the pair ahead of and behind it (Fig. 1). The array is fed from the forward end of the feed line (where the elements are the shortest), with a 4:1 balun and RG-8 coaxial cable running back to the shack. A review of an earlier article by YV5DLT is helpful.

When I began assembling materials for construction of the antenna, fiberglass vaulting poles were hard to find and expensive. I considered plastic PVC water pipe, but that thought was soon abandoned as it was too limber. A local sporting goods store had a stock of fishing rod blanks, but none were strong or long enough (at least 15 feet). Tapered rods are also hard to splice together. Finally, I located a plastics supplier who stocked solid fiberglass rods 1/2 inch, 3/4 inch, and 1 inch in diameter and 6 feet long.\2 The 3/4-inch diameter rods were selected as the best trade-off between stiffness, weight and cost. These rods came without surface covering and after a few hours of handling, my hands itched from tiny glass slivers. The supplier advised that an aerosol spray of clear plastic should eliminate this problem and minimize the weathering that I have noted with time.

I chose a design that would use five fiberglass

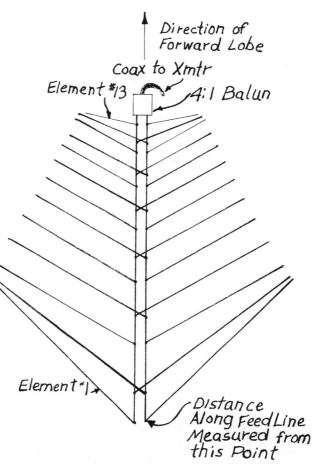

Fig. 1 — Connection of radiators. Alternate pairs are transposed as they connect to the feed line.

spreaders fastened at a common hub for the support construction, rather than the four vaulting poles and crossbow employed by Eckols. See Fig. 2. The hub was made from a 6-inch square, 1/4-inch thick steel plate to which a local welder attached five pieces of 3/4-inch pipe at the appropriate angles as shown in Fig. 3. A 1-1/4-inch pipe flange was mounted on the underside of the plate with four no. 1/4-20 screws. To assemble the antenna, I found a flat spot in the yard where I could position a section of 1-1/4-inch pipe on the ground to place the hub about waist height. It was easy to assemble the various parts this way, and I could cut and stretch the antenna elements close to the ground. After completion, the array was tested and operated on the air, before being hoisted into position on the roof.

The ends of the first lengths of fiberglass were wrapped with plastic electrical tape for about 8 inches and pushed as far into the pipes at the hub as they would go. Two holes were then drilled about four inches apart and at right angles to each other for placement of the screws that secure the rods to

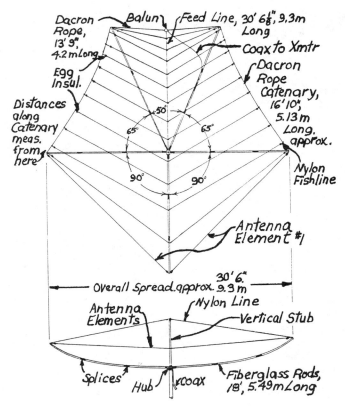

Fig. 2 — Layout of fiberglass rods and catenaries. The top view is the plan view, while the bottom shows the elevation view.

Fig. 3 — Construction of central hub. Weld five sections of black pipe to a flat steel plate; attach a pipe flange to the bottom to hold the pipe support and another to the top to support the vertical stub.

each pipe.

To obtain the necessary spreader length, three 6-foot sections of rod were spliced end to end. My first method proved almost disastrous. I had slipped the ends into a section of aluminum tubing with a 0.055-inch wall thickness. The splice looked strong enough, but as the completed antenna was being transported on to the roof, one of the splices rup-

tured and the whole antenna collapsed. I had exceeded the yield strength of aluminum. After reviewing books on material strengths, my next attempt proved strong enough to weather our severest storms. One-foot sections of 1-inch aluminum angle with an 1/8-inch thickness were more than adequate. Two holes were drilled at right angles through each rod and angle, and secured with no. 8-32 steel screws to make a solid splice.

At the outer ends of the rods, a tie point was necessary to attach the end of the feed line at the rear of the array, and to secure the ends of the antenna elements and tie lines to the other four rods. PVC pipe caps were too fragile. At the local hardware store, I purchased bronze straps that plumbers use to secure 3/4-inch copper pipe. The straps were mounted at the end of each rod by drilling through the strap and rod, and securing it with a no. 8-32 screw. A shim of copper pipe between the rod and strap made a better fit. See Fig. 4A.

Next, I constructed 13 feed-line insulators made of 2-inch square, 1/2-inch thick plastic acrylic. Two holes provided the spacing for the feed line (Z_o = 378 ohms before loading by the antenna elements), and two additional holes hold the ends of the antenna elements. Spacing of the feed-line holes is critical and should be drilled with care. See Fig. 4B.

The following step was to construct the feed line. Beginning at the rear of the array, the two feeder wires were attached to the rear fiberglass rod with egg insulators (Fig. 5). The two conductors were then threaded through holes in the 13 plastic insulators. These were spaced at the precise position for each pair of antenna elements to connect to the feed line. For the first antenna, I used insulated, hard-drawn copper wire and removed the insulation only at the connection points for the antenna elements. For the other antenna, Copperweld wire was used with sleeves for insulation except at the connection points. The two methods seemed to work equally well.

Spacing of the elements is critical. Distances were measured from the rear end of the feed line to within 1/8 inch of the values shown in Table 1 to maintain a low SWR. Remember that the longest radiators are connected directly to the feed lines, but the next shorter radiators must be transposed when connected to the feed line. Then the following pair are again connected directly and so on down to the shortest elements. To measure the antenna lengths accurately, I found it easier to insert the loose end of the wire through the correct hole of the feed-line insulator, then secure and solder it to the appropriate feed-line wire. After that I cut the radiator wire to size, measuring the length from the point it left the feed-line soldered joint. See Table 1. At the other end I allowed an extra few inches of wire to secure it to an egg insulator. Then a length of monofilament nylon fish line was tied with a bowline knot through the other hole of the insulator and left dangling until all of the antenna elements were attached to the feed line and cut to length.

Finally, the feed line was stretched taut by securing its front end to the center of a rope attached to the two front fiberglass rods. At this point, the array began to take shape. The two ends of the longest radiators were connected to the ends of the two side rods; dacron ropes run from these rods to the two front rods. These last two rope

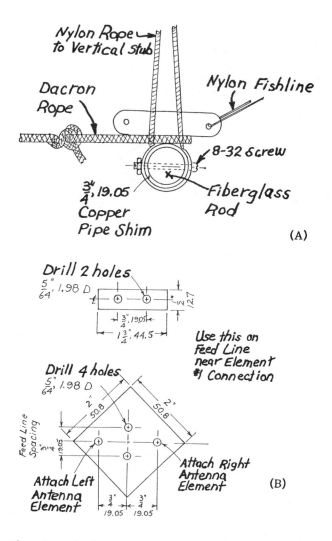

Fig. 4 -- A shows straps attached to ends of fiberglass rods made from 3/4-inch bronze waterpipe straps. To prevent cutting the rope and fish lines, file off sharp edges and ream holes. At B, feed-line insulators made from acrylic plastic 1/2 inch thick. Note: At element no. 1, feed-line wires lie side by side. At element no. 2, they twist so that they are one above the other for the connection of the rest of the elements. At the balun, the wires are again side by side.

catenaries now complete the perimeter of a five-sided figure to which the rest of the antenna elements could be attached. It is better to allow at least 10 feet of extra rope as slack while the fiberglass rods are slowly bent upward into the desired shape. The bow provides the necessary tension on the dacron ropes and antenna elements to keep them taut.

When I first constructed the array, I used braided polypropylene swimming pool rope. After less than a year, the ultraviolet rays from the sun had reduced the rope to short brittle needles with little tensile strength. A replacement with 3/16-inch dacron rope has weathered two winters and seems as strong as ever.

An optional vertical stub is mentioned in Eckols' article. It extended upward from the central hub with ropes going out to the ends of each of the rods. Without this setup, my array was just as stable with the rods bent in an upward arc as with a downward arc. But, like the proverbial flip-flop, it was prone to flip from its upward position to the downward one with little notice.

The next thing I installed was a post of 3/4-inch pipe, 6-feet in length, threaded into a flange mounted on top of the hub. A section of PVC plastic was slipped over the pipe to keep the feed line from shorting if the two came in contact. At the top of the pipe a 3/4-inch to 1/2-inch reducer, a short nipple and a 1/2-inch pipe cap finished it off. Nylon lines could now be looped over this cap and the ends secured to each of the rods. This helped to keep the proper bow in the rods as the elements were being secured to the dacron rope catenaries. You may find additional lines from the midsection of the rods to the post helpful. When my nylon lines need replacing, I can slip the old lines off and replace them with the aid of a long pole without having to climb on top of the array to untie knots.

The rest of the radiators can now be stretched into place by attaching the ends of the fishing line that were previously left dangling. I found it helpful to indicate the positions where the fish lines terminate along the catenaries with a marking pen or masking tape. Then, using an ice pick, I poked holes through the dacron rope, pulled the fish line taut, and tied it with a temporary knot. Beginning with the longer radiators, you can work your way toward the front of the array, connecting first one radiator on the right to its catenary and then one on the left. After all the radiators have been stretched into position, some may be too taut and others too lose. Don't give up, you are almost there.

The next attempt to equalize the tension on each radiator should be undertaken only when you feel relaxed and have an afternoon ahead of you. With the help of a friend, tighten the loose radiators and lengthen those that appear to be too taut. Don't be afraid to adjust the length of the catenary ropes. Small changes in the catenaries will provide more or less bend to the spreaders, and thus, impart more or less tension on the catenaries. Finally, you can make the appropriate adjustments on the individual fish lines. Don't expect the wire elements to ever be real taut; it just isn't possible with this array. Balancing the array is important to reduce wear in the rotator bearings. My array was heavier at the front. This was offset by wrapping a sheet of lead around the rear rod at its outer end for a counterbalance.

At this point it's time to connect the coax to the rig and begin making SWR measurements on the frequencies you will operate on. I found that although the readings were acceptable with the array on the ground, some improvement could be expected once the beam was properly positioned in the air. Table 2 gives a comparison of some measurements I made with the two antennas.

The final step is installation. The array was lifted intact to an antenna rotator already mounted on the peak of my roof. A gin pole was temporarily placed along side the rotator mount. Then a rope tied to the central hub was threaded through a pulley atop the gin pole. One man helped guide the fiberglass rods while climbing an extension ladder; two other fellows on the roof tilted the array into its horizontal position while the array was supported by the rope. Finally the rope was slacked off to allow the 1-1/4-inch pipe to drop into the top of the rotor and be tightened in place.\3

I have had many pleasurable hours of DXing with my Spiderweb array. I hope you will share the same success.

References

\1 Eckols, Ansyl, YV5DLT, "The Telerana — A Broadband 13- to 30-MHz Directional Antenna," **QST**, July 1981, p. 24 to 27.

\2 TAP Plastics, Inc., 3011 Alvarado St., San Leandro, CA 94577, (415) 357-3755.

\3 A coat of paint suitable for metal surfaces should help the weathering. The vertical stub makes a good lightning rod; don't forget to bring a grounding wire back to earth. I used no. 8 wire back to a driven ground rod.

Fig. 5 — Detail showing connection of antenna element no. 1 to rear end of feed line. Note the egg insulators attached to the strap shown in Fig. 4A. Vertical rope from end of fiberglass rod was used to balance array during construction. It was replaced by weight of lead sheet wrapped around rod at this end.

Parts List

Qty	Description
15	Fiberglass rods, solid, 3/4 inch diameter, 6 feet long
10	Aluminum angles, 1 inch x 1 inch x 1/8 inch section thickness, 12 inch long
5	Straps for 3/4-inch copper water pipe, see Fig. 4A
5	Black gas pipe, 3/4 inch diameter, 8 inch long, no threads needed
1	Steel plate, 6 inch x 6 inch x 1/4 inch thick
1	Pipe flange, 1-1/4-inch thread
1	Pipe flange, 3/4-inch thread
1	Galvanized pipe, 1-1/4 inch diameter, 4 feet long, threaded one end, to support antenna during construction. Later it can be shortened (or lengthened) to fit into antenna rotator.
1	Galvanized pipe, 3/4 inch diameter, 6 feet long, threaded both ends
1	Plastic pipe, PVC 1 inch diameter, 1/16 inch wall, 6 feet long, to fit over 3/4-inch pipe
1	Galvanized pipe reducer, 1/2 inch to 3/4 inch
1	Pipe nipple, 1/2 inch, 2 inches long
1	Galvanized pipe cap, 1/2 inch
65 ft	Insulated no. 14 copper wire for feed line, see text
13	Acrylic plastic, 2 inch x 2 inch x 1/2 inch thick for feed line insulators
1	Acrylic plastic, 1-3/4 inch x 1/2 inch x 1/2 inch thick for spacer at element no. 1
1	Balun 4:1 W2AU or equiv. RG-8 coax and connectors to feed from balun to transmitter
325 ft	Stranded copper wire, no. 14 for antenna radiators
28	Small egg insulators
50 ft	Monofilament nylon fish line, at least 50 lb test
70 ft	Dacron rope, 3/16 inch diameter
125 ft	Nylon or dacron rope, 1/8 inch diameter
60	Steel, RH machine screws, no. 8-32, 1-1/2 inch long, also nuts flat washers and lock washers
4	Steel, FH machine screws, no. 1/4-20, 1/2 inch long to fasten 1 1/4-inch pipe flange to plate
4	Steel, FH machine screws, no. 8-32, 1-1/2 inch long to fasten 3/4-inch pipe flange to plate

TABLE 1

ANTENNA RADIATOR LENGTHS AND SPACINGS

Radiator Number	Radiator Lengths		Connection Points along Feed Line Measured from Rad #1		Distance along Catenaries Measured from Rad #1	
	feet - in	meters	feet - in	meters	feet - in	meters
1	20 - 4-1/8	6.20	--	--	--	--
2	18 - 3-11/16	5.58	4 - 0-13/16	1.24	--	--
3	16 - 5-11/16	5.02	7 - 10-1/2	2.40	1 - 3-3/4	0.40
4	14 - 9-15/16	4.52	11 - 1-7/8	3.40	4 - 1-3/16	1.25
5	13 - 4-1/8	4.07	14 - 1-11/16	4.31	6 - 8-11/16	2.05
6	12 - 0-1/8	3.66	16 - 9-9/16	5.12	9 - 0-11/16	2.76
7	10 - 9-3/4	3.29	19 - 2-5/16	5.85	11 - 1-1/16	3.38
8	9 - 8-3/4	2.97	21 - 4-5/16	6.51	13 - 0-5/16	3.97
9	8 - 9-1/16	2.67	23 - 3-15/16	7.11	14 - 8-3/8	4.48
10	7 - 10-9/16	2.40	25 - 0-13/16	7.64	16 - 1-11/16	4.92
11	7 - 1-1/8	2.16	26 - 7-11/16	8.12	16 - 6-13/16	5.05
12	6 - 4-5/8	1.95	28 - 1	8.56	16 - 10	5.13
13	5 - 8-15/16	1.75	29 - 4-3/8	8.95	16 - 10	5.13
Balun connection			30 - 6-1/8	9.30	--	--

TABLE 2

TYPICAL VSWR READINGS

Frequency MHz	First antenna 25 feet in air	Second antenna approx. 6 feet above ground
14.050	1.35	1.1
14.150	1.3	1.2
14.300	1.2	1.45
21.050	1.75	1.8
21.150	1.6	1.6
21.300	1.3	1.7
21.445	1.1	1.8
28.050	2.3	2.2
28.150	1.8	2.1
28.300	1.35	1.75
28.450	1.1	1.45
28.600	1.05	1.3
28.750	1.15	1.2
28.900	1.45	1.25

Antennas fed with Kenwood TS-820 and MFJ Versa Tuner II, Model 949 B

in Direct Coax position.

Other Beam Antennas

A Simple Log-Yag Array for 50 MHz

By John J. Meyer,* N5JM

*112 Sherwood Forest Drive, New Orleans, LA 70119

The 50-MHz band is a playground of sorts for the antenna builder. It offers many interesting possibilities such as high gain, small physical size and the potential of making DX contacts with a homemade antenna and low power. Interestingly, at 50 MHz and higher, antenna construction becomes simpler and within the realm of the novice builder. Hand tools can be employed in constructing many antenna designs for this frequency range and since most construction materials for VHF antennas can be purchased at a hardware store, the builders job is easier and inexpensive. This antenna makes an ideal weekend project for the amateur who would like to build a beam at a fraction of the cost of a "store bought" Yagi.

Most 50-MHz antennas are Yagis. They offer high gain for their size, yet there are lesser known antenna variations offering benefits such as greater bandwidth, ease of matching and higher gain for the same boom length. The Log-Yag Array is such an antenna.

The Log-Yag Array, popularized by Oliver Swan in the 1960s, combines the bandwidth of the log-periodic antenna with the high gain of a Yagi.[1] Those interested in the theory and design should reference the articles by K4EWG, W4BBP, W4AEO and W6SAI for an in-depth explanation on how they work.[2,3,4] In essence, the Log-Yag has two or more dipole-driven elements. Each is fed 180 degrees apart, and consists of one or more parasitic Yagi elements to both sharpen the beam pattern and increase the front-to-back ratio. Element spacing tends to be closer than a Yagi, and element lengths are also slightly longer.

This article describes a basic design that can be duplicated easily. The boom is a 10-ft section of TV mast from Radio Shack (or aluminum, if available) and the elements are 1/2-inch EMT (electricians metallic tubing) from the local hardware store. Dimensions and spacing are not critical, and good construction techniques should be followed, just as you would with any antenna. Moisture and corrosion are serious problems so care should be taken to paint and protect the driven-element supports made of wood and all the associated hardware. Element-to-boom mounting is best accomplished as illustrated in **The ARRL Handbook.**[5]

The elements are cut to the following lengths:

Reflector: 116 "
Rear Driven Element: 54 " each length (111 " tip to tip, 3 " center)
Front Driven Element: 53 " each length (109 " tip to tip, 3 " center)
First Director: 107 "
Second Director: 104 "

A complete diagram of the 50-MHz Log-Yag Array can be seen in Fig. 1.

The driven elements have a 3-inch center spacing. The elements should be mounted on the 10-ft boom as follows:

Reflector to Rear-Driven Element: 23 "
RDE to Front-Driven Element: 26 "
FDE to First Director: 19 "
FD to Second Director: 50 "

The driven elements are insulated from the boom with wood, while the other parasitic elements are mounted directly to the boom. A copper or aluminum feed line is transposed between the two driven elements to achieve the 180 degree phase shift. The Log-Yag Array is fed at the front driven element with a 52-ohm balanced feed and exhibits an impedance of about 60 ohms. The 1:1 balun in **The ARRL Handbook** works quite well for this purpose. It should be noted that the balun must hang free from the antenna and should not be taped or otherwise attached to the antenna, or the results will be poor radiation patterns and strange SWR ratios. There is little else in the construction that will affect the SWR as much as an improperly built or placed balun.

Moving the first director about on the boom and changing its length affects both the SWR and impedance. Should the builder wish to vary these values, this would be the place to start. To a lesser extent, the reflector also controls these values. Experimentation can be done by placing the reflector element on a wood or other non-metallic support and pointing the antenna straight up in the air (away from any wires!) so that it can be worked on at ground level.

This Log-Yag Array is flat across the first megahertz of the 6-meter band with an SWR of less than 1.3:1 and is usable up to 52 MHz. The front-to-back ratio is better than 25 dB and it has a side null in excess of 35 dB. The front lobe is broad and the antenna can be rotated 45 degrees off a station without any appreciable loss of signal. There are no side lobes. Increasing the boom length and adding additional directors will sharpen the front lobe for those who want greater directivity. Although no comparisons have been made with larger antennas, I believe this antenna will outperform a Yagi of the same boom length. Good DX!

References

[1] "The Swan Multidrive 2-Meter Antenna," **QST**, p. 42, Oct. 1969.
[2] Rhodes and Painter, "The Log-Yag Array," **The ARRL Antenna Anthology,** pp. 49 to 52.
[3] Smith, numerous articles in **Ham Radio** since 1974.
[4] Orr, **Radio Handbook,** 22nd edition, pp. 30-15 to 30-17.
[5] **The ARRL Handbook for the Radio Amateur,** 62nd edition, (ARRL: Newington), 1984.

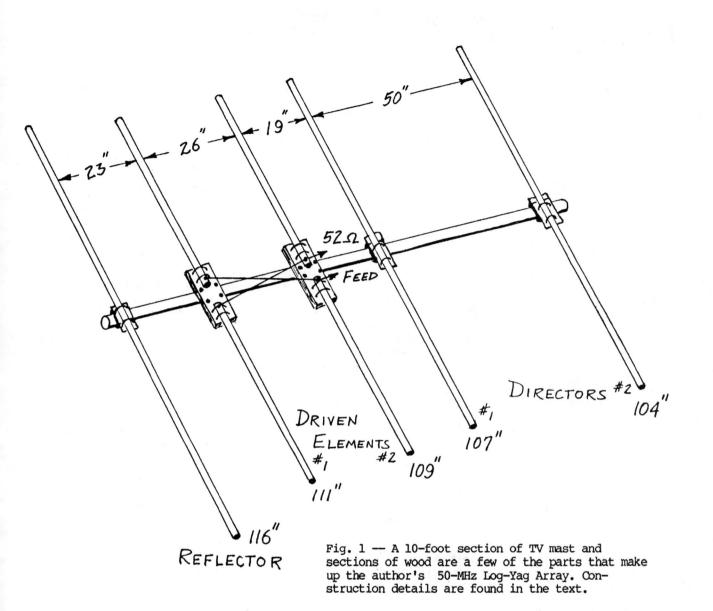

23" 26" 19" 50"

52 Ω

FEED

REFLECTOR 116"

DRIVEN ELEMENTS #1 111" #2 109"

DIRECTORS #1 107" #2 104"

Fig. 1 — A 10-foot section of TV mast and sections of wood are a few of the parts that make up the author's 50-MHz Log-Yag Array. Construction details are found in the text.

Designing X-Beams

By Brice Anderson,* W9PNE

*P. O. Box 14, Lancaster, IL 62855

The X-Beam is a high-performance broadband antenna that is ideal for radio operators with limited space. If you are searching for an understanding of this compact but powerful antenna, the explanation and formulas found in this article should put it on solid ground for you.

Description

The X-Beam is a compact version of a 2-element Yagi whose performance equals or exceeds that of a Yagi. It has good bandwidth, a 6-dB forward gain over a dipole, and a 15- to 18-dB front-to-back ratio. The formulas presented will enable you to calculate X-Beam arm and tail lengths for any frequency.

The X-Beam name is derived from its shape, shown in Fig. 1. It consists of four arms often built from aluminum tubing and four loading wire tails. As there is no boom, the antenna is usually mounted in a horizontal plane, though a vertical-plane position can be used. It is considerably smaller and lighter than a 2-element Yagi for the same band.

Construction

A square piece of heavy plywood, treated for weather protection, makes a suitable hub for 15- and 20-meter X-Beams. The arms are secured to the hub with pipe brackets, and a nylon cord is strung tightly around the beam from tip to tip. This procedure strengthens the assembly and supports the tails.

You could use an alternative method of construction with two pieces of aluminum angle stock bolted together as the X hub. The arms may then be clamped to the hub with screw type hose clamps, using plastic tubing to insulate the driven arms.

Versions

Formulas are given for full-size and minimum-size X-Beams. The full-size beam provides a close match to 50-ohm coax and the greatest possible bandwidth. It is recommended for 10, 15 and 20 meters. The minimum-size beam has similar performance, but the SWR is approximately 1.4:1. Because of the shorter arms, the ends of the tails nearly touch. It is used for 30 and 40 meters.

Many measurements on a large number of experimental X-Beams were used to develop the formulas. The arms were made of tubing with a length-to-diameter ratio of about 200:1. Stranded no. 19 vinyl-covered wire was used for the tails. The use of larger tubing or larger wire size will require slightly shorter tails than given by the formulas, and vice versa. Arms may be made of wire supported by nonconductive materials such as wood or bamboo. In this case, the tails will have to be made considerably longer.

Formulas

For Full-Size:

Arm length (ft) = 195/f (MHz)   (each arm, about 82% of half a Yagi driven element)

Total driven element (ft) = 603/f (MHz)   (about 27% greater than a Yagi driven element)

Driven element tail (ft) = 106.5/f (MHz)   (each tail)

Total director element (ft) = 575/f (MHz)   (4.65% shorter than driven element)

Director tail (ft) = 92.5/f (MHz)   (each tail)

Do not use arms much longer than calculated by the full-size formula. As the arms get longer and the tails get shorter, the pattern changes gradually to four broad lobes.

For Minimum-Size:

Arm length (ft) = 177/f (MHz)   (each arm, about 75% of half a Yagi driven element)

Total driven element (ft) = 603/f (MHz)   (same as for full size)

Driven element tail (ft) = 125/f (MHz)   (each tail)

Total director element (ft) = 575/f (MHz)   (4.65% shorter than driven element)

Director tail (ft) = 110.5/f (MHz)   (each tail)

General

A purchased beam is shipped unassembled, but it comes with a table of the exact element lengths and spacings for CW or phone frequencies. Since it is a replica of the manufacturer's model, it will perform well once it is constructed. A homemade beam, on the other hand, will not be like the one described here. It will probably have to be tuned to resonance. Luckily, this procedure is simple.

Tuning and Adjustment

The tails should be made a few inches longer than the calculated lengths and the director tails shorter than the driven element tails as shown in the tables. For tuning, the beam should be mounted on a pole at least 10 ft in height. To determine resonance, an SWR bridge is placed in the coax feed line. An alternative method is to couple a dipper to the director element.

Using 5 W of power or less, quickly make some SWR measurements starting at the low-frequency end

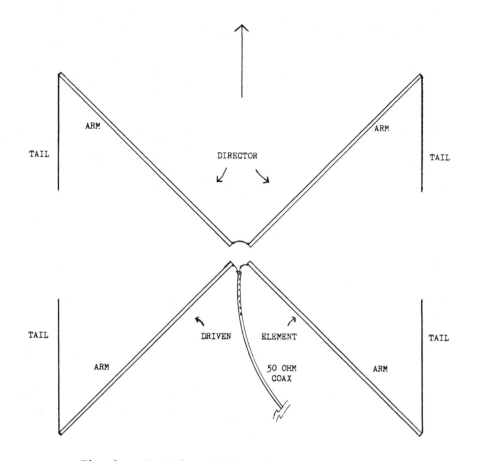

Fig. 1 — Top view of the X-Beam. The arms are made of tubing. Nylon cord is strung tightly around the beam from tip to tip, to strengthen the assembly and support the wire tails.

of the band. Do not cause QRM! The lowest SWR occurs at the resonant frequency. If the resonant frequency is too low, cut one inch from each of the four tails (1/2 inch on 10 meters). Again determine resonance. Repeat this process until the antenna resonates at the desired frequency. The difference in tail length must be as shown in the tables. The resonant frequency will increase somewhat when the antenna is mounted on the tower.

Baluns

The radiation pattern is slightly unsymmetrical because of currents on the coax shield. While this scarcely effects the antenna performance, purists may connect a 1:1 balun to the driven element. For a balun, the coax can be coiled into a 6-in inside diameter coil, taped together and attached to the mast under the beam. The coil should be 12 turns for 15 and 20 meters and 6 turns for 10 meters.

Conclusion

The actual peformance of the simple X-Beam is amazingly similar to that of a 3-element Yagi. Put one up and challenge the "big guns."

Table of Dimensions for Full-Size X-Beams

	Arm		Driven El. Tail		Director Tail		Tail Diff.	
MHz	Ft	In	Ft	In	Ft	In	Ft	In
7.050	27	8	15	1 1/2	13	1 1/2	2	
10.125	19	3	10	6	9	1 1/2	1	4 1/2
14.100	13	10	7	6 1/2	6	6 1/2	1	
21.100	9	3	5	1/2	4	4 1/2		8
28.200	6	11	3	9 1/2	3	3 1/2		6

Table of Dimensions for Minimum-Size X-Beams

MHz	Arm		Driven El. Tail		Director Tail		Tail Diff.	
	Ft	In	Ft	In	Ft	In	Ft	In
7.050	25	1 1/2	17	9	15	9	2	
10.125	17	6	12	4	10	11	1	5
14.100	12	6 1/2	8	10	7	10	1	
21.100	8	4 1/2	5	11	5	3	8	
28.200	6	3 1/2	4	5	3	11	6	

An HF Phased Array Using Twisted-Wire Hybrid Directional Couplers

By James V. Melody,* WA2NPJ/KX6JM

*145 South Zurich Ave., Egg Harbor City, NJ 08215

Many articles have been published on phased arrays for 40 and 80 meters.[1] The advantage of a phased array over a beam is that it can be turned electronically, with no need for large, expensive electro-mechanical rotators. In addition, the beam can be electronically aimed at a station during transmit, with an interferring station switched to a null on receive.

Several combinations of antennas at various phases have been simulated and tested by experimenters. From their data, it seems that the most practical phased array for 40 and 80 meters consists of four vertical elements arranged in a square and spaced one-quarter wavelength apart. These four elements are fed with equal power at phases shown in Fig. 1. Such a configuration will give a gain of 5.3 dB, a front-to-back ratio of 25 dB, and a front-to-side ratio of 12 dB.[2] Previous designs of HF phased arrays have used Wilkinson power dividers or phasing lines, both of which employ long lengths of coaxial cable to obtain the necessary phase change and power split.[3,4]

In this design, a series of twisted-wire quadrature hybrid directional couplers are used to obtain the power split and phase shift. A circuit diagram of a single twisted-wire hybrid is shown in Fig. 2. A single hybrid divides the power input equally between ports 2 and 4 at port 1, with no power entering port 3 if all $Z\phi$ are equal. The hybrid also shifts the phase a total of 90 degrees between ports 2 and 4. If these hybrids are cascaded as shown in the block diagram in Fig. 3, four outputs are obtained. Each of the four outputs are of equal power and two have zero degrees phase, one -90 degrees, and one +90 degrees. This is the exact power split and phase shift needed for our antenna array. The complete circuit is shown in Fig. 4.

The makeup of such a network begins with the design of the twisted-wire hybrid.[5,6] The coil, L1, shown in Fig. 2 is a bifilar winding on a toroidal core capable of handling the expected power from the transmitter. In Fig. 5A, the inductance of this twisted pair is given by:

$$L = \frac{Z\phi}{2\pi f\phi}$$

The frequency of interest is the middle of the 40-meter band and the impedance is 50 ohms.

$$L = \frac{50}{2\pi \times 7.15 \text{ MHz}} = 1.12 \ \mu H$$

By an examination of catalogs and application notes, I decided to use an iron-powder toroid.[7] I picked the "mix 2" with permeability 10 because it covers the 7-MHz band well, and it is readily available. The number of turns required for an iron-powder toroid is given by:

$$T = 100 \left[L(\mu H)/AL \right]^{1/2} \qquad \text{(Eq. 1)}$$

where AL is a parameter of the toroid given by the distributor.

Since the coil will be dealing only with an ac voltage, a core large enough to handle the maximum ac flux density is required. The maximum ac flux density gauss (G) is given by:

$$Bmax = \frac{ERMS \times 10^8}{4.44 \ AeNf\phi} \qquad \text{(Eq. 2)}$$

where

Ae = effective cross-sectional area (cm^2)
$ERMS$ = the applied voltage
$f\phi$ = the frequency of interest
N = the number of turns

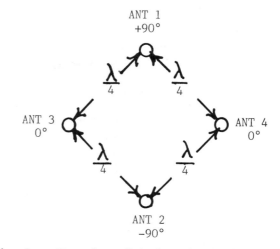

Fig. 1 — Top view of 4-element square array showing phases and relative power at each antenna.

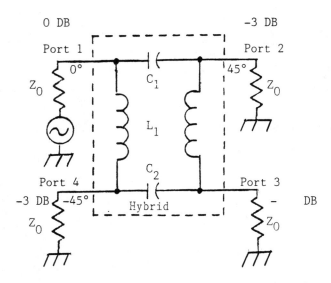

Fig. 2 — Circuit diagram of a single twisted-wire hybrid. All Z_ϕ must be equal, and C1 equals C2.

In addition to the above Eqs. 1 and 2, there is a general rule to follow for choosing the largest core: Be consistent with the price and space available. Based on this data, a T-200-2 core was selected. The following are the parameters for a T-200-2 supplied by the manufacturer:

$$AL = 120 \ \mu H/100 \text{ turns}$$

$$Ae = 1.33 \text{ cm}^2 = \text{effective area}$$

Substituting the required 1.12 μH inductance and the value for AL into Eq. 1, the number of turns is:

$$F = 100(1.12/120) \ 1/2 \quad = 9.6 \text{ turns}$$

Assuming that we now have 1000 W RMS into a 50-ohm load, Bmax from Eq. 2 is:

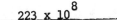

$$Bmax = \frac{223 \times 10^8}{4.44 \times 1.33 \times 9.6 \times 7.15 \times 10^6}$$

$$Bmax = 54.6 \text{ gauss}$$

From the manufacturer's specification sheet, type 2 material will support 14 kG. The ten turns of no. 18 wire will easily fit on the core. Therefore, the T-200-2 is an acceptable choice.

If similar calculations are performed for a T-130-2 core at 500 W RMS (3 dB down), ten turns are required and the maximum gauss is 53.5. Thus, the T-130-2 core is a good selection for the second and third hybrid.

Having chosen the core and the turns for L1, the capacitors must now be determined. In Fig. 5B, the capacitance required is given by:

$$CReq = \frac{1}{2 \pi f\phi Z\phi}$$

$$Z\phi = 50 \text{ ohms}$$

$$f\phi = 7.15 \text{ MHz}$$

Therefore, CReq is 445 pF. This value includes both C1 and C2 as well as the capacitance between the twisted bifilar turns of L1.

To implement the three hybrids, first wind the toroids with the twisted pair. No. 18 wire with Teflon covering is a good choice. Twist the wires together with about six twists to the inch. Wind all three toroids with about ten turns. Here, it is convenient to hold the wires in place with two plastic tie wraps. Connect the wires as shown in Fig. 5A and measure the inductance of the coil. (See **The ARRL Handbook for the Radio Amateur**.) It should be 1.12 μH. If not, remove or add turns to obtain a value as close as possible to this number. Once this is done, tighten the tie wraps to hold the wires firmly in place. Recheck the inductance and spread or compress the windings to decrease or increase the inductance, respectively, to get it to the exact value desired. In actual tests, the cores shown in

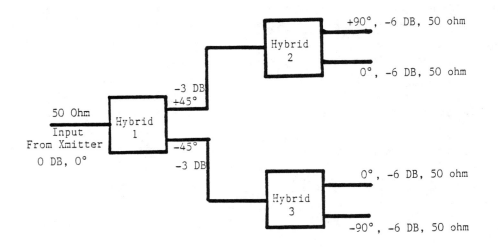

Fig. 3 — A block diagram of three twisted-wire hybrids stacked to give four outputs of equal power and correct phase for a 4-element phased array.

Fig. 6 required 7.5 turns spaced over about 80% of the core. Although you purchase parts that look identical, you will still need to measure the inductance because of variations in manufacturing.

Now measure the capacitance (Cbifilar) between the wires of the twisted pair. This should be very small — about 30 pF. Since this capacitance is part of the capacitance between the halves of the hybrid (Fig. 5B), each capacitor in the hybrid must be:

$$C1 = C2 = \frac{CReq - Cbifilar}{2}$$

For hybrid 1, Cbifilar is 32 pF, and for hybrids 2 and 3, Cbifilar is 28 pF. These values must be measured for your unit if other than no. 18 Teflon-covered wire is used.

The capacitors required for each hybrid are therefore:

Hybrid 1: $\frac{445 - 32}{2}$ = 206 pF = C1 = C2

Hybrids 2 and 3: $\frac{445 - 28}{2}$ = 208 pF = C3 = C4 = C5 = C6

Select a parallel set of capacitors to obtain values close to this range. This, of course, must be done by measuring individual capacitors since they can vary from marked values.

We now have the three hybrids of our phase shifter. Looking at Fig. 4, we can see how the three hybrids are connected into the final circuit for a switchable phased array for 40 meters.

The last major design problem is the dummy load at port 3 of each hybrid. As stated earlier, the power into port 3 is zero if all $Z\phi$ are equal. This is not the case in actual practice since no antenna will maintain a 1:1 SWR across the entire band. By experiment, it was found that less than 10 W of power went into port 3 of hybrid 1 if all four verticals had an SWR of less than 1.5:1. These tests were run with an SB-200 transmitter which was putting out about 600 W. It was therefore determined

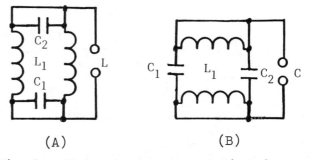
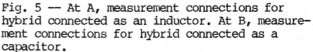

(A) (B)

Fig. 5 — At A, measurement connections for hybrid connected as an inductor. At B, measurement connections for hybrid connected as a capacitor.

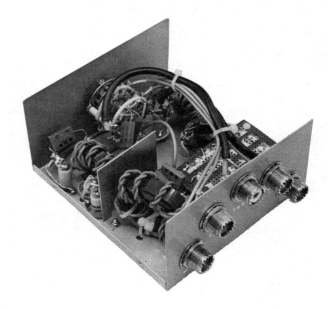

Fig. 6 — The completed phase shifter. (Photo by Jim Wedge)

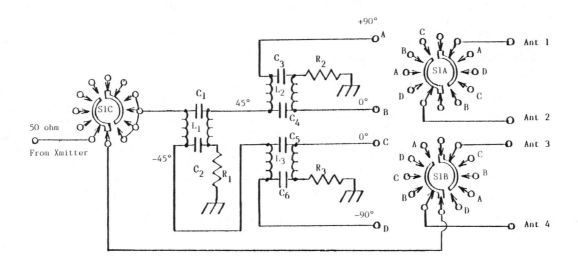

Fig. 4 — Complete circuit diagram for a 4-element switchable phase shifter for a 40-meter phased array.

that a 20-W noninductive resistor was needed for R1, and 10-W resistors for R2 and R3. These were constructed of parallel 2-W resistors. The value should be measured and resistors added or deleted to get as close to 50 ohms as possible. For R1 it was found that eleven 470-ohm, 2-W, 5% resistors gave almost 50 ohms exactly, while R2 and R3 consist of six 300-ohm and one 470-ohm resistors in parallel. Check your values before you solder them together.

The finished product is shown in Fig. 6. Since toroids are used, the system can be extremely compact and internal shielding can be kept to a minimum. One switch position is used to feed one vertical directly. This was to give a "nondirectional" position for testing and listening purposes. If you plan to use one of the larger kW rigs or will not be able to get all your verticals close to 1.5:1 over the band, you should use higher wattage resistors for R1, R2, and R3. There are some important points to remember. All lines from the phase shifter to the antennas must be the same length and have the same velocity factor. All antennas must be fed the same way. Different loading schemes at each antenna will, of course, change the phase.

Four one-quarter wavelength verticals were spaced at four sides of my home. Two are wires running up trees and the other two are the remains of a TH6DXX. See Fig. 7 for an example of the antenna. I installed an 8-ft ground rod at each antenna since I have no room for radials. Also, because of space limitations, the distance between antennas is not exactly one-quarter wavelength nor exactly equal. Still, the gain closely approximates the theoretical value, as do the front-to-back and front-to-side ratios. Look in **The Handbook, The ARRL Antenna Anthology,** and **The ARRL Antenna Book** for a complete discussion of verticals and ground systems which could be used at your location.

[Editor's Note: Fig. 8 shows the calculated radiation pattern of this array.]

References

\1 Anderson, Marian, WB1FSB, **The ARRL Antenna Anthology,** 1978, pp. 114-122.
\2 Atchley, D. W., Stinehelfer, H. E., White, J. F., "360 Steerable Vertical Phased Arrays," **QST,** April 1976, p. 27.
\3 Wilkinson, Ernest J., "An N-Way Hybrid Power Divider," **IRE Transactions on Microwave Theory and Techniques,** January 1960, p. 116.
\4 Atchley, D. W., "A Switchable Four-Element 80-Meter Phased Array," **QST,** March 1965, p. 48.
\5 Fisher, R. E., "Broad-band Twisted-Wire Quadrature Hybrids," **IEEE Transactions on Microwave Theory and Techniques,** May 1973, p. 355.
\6 Fisher, Reed, "Twisted-Wire Quadrature Hybrid Directional Couplers," **QST,** January 1978, p. 21.
\7 **Application Notes,** Amidon Associates.

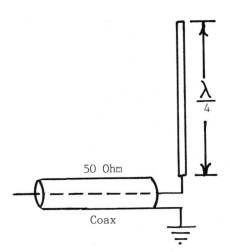

Fig. 7 — Possible antenna and feed for each element of phased array.

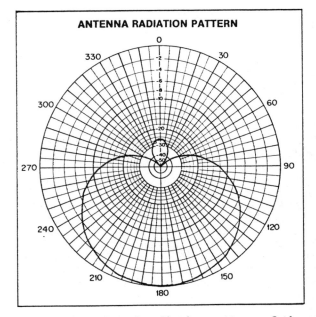

Fig. 8 — Calculated radiation pattern of the array shown in Fig. 1, for a zero-degree radiation angle. A perfect earth and equal currents in the four elements are assumed.

Parts List

L1:	T-200-2 Amidon iron-powder toroid
L2 and L3:	T-130-2 Amidon iron-powder toroid
R1:	Ten 470-ohm, 2-W carbon resistors
R2 and R3:	Six 300-ohm, 2-W carbon resistors
C1 thru C6:	See text.
Switch:	6 pole, 5 position (need only 5 pole) Centralab no. PA-2021
Metal Utility Cabinet:	5-1/4 x 3 x 5-7/8 inch, Radio Shack no. 270-253

Two views of the phase-shifter-network box. In the nondirectional switch position (ND), only Ant. 4 is fed. (Photos by Jim Wedge)

LARAE—Line Array of Rotary Antennas in Echelon

By A. J. F. Clement ,* W6KPC

*Western Science Foundation, Rt. 1, Box 116,
McFarland, CA 93250

As an ardent and devoted DXer, I have built, tested and operated four large antenna arrays. Each was completed near the peak of successive sunspot cycles, i.e., 1948, 1958, 1969 and 1979.[1,2,3,4] These high-performance antenna arrays were always directed toward better performance on the 10-, 15- and 20-meter bands. During 1975 I decided to design a high-performance array-type antenna that could be adapted easily to both the low-frequency and high-frequency bands. As this article is being prepared, LARAE is being constructed on my antenna farm for the 75-meter band.

The scheme offered here is intended to offer a much larger antenna aperture than can be realized from even a long-boom standard Yagi antenna. This array will also reduce QRM and QRN considerably, since it will have a very narrow beamwidth in azimuth. LARAE's angle of take-off (vertical angle of radiation) will depend, principally, on its height above the earth, but not entirely so. Height consideration alone would indicate a vertical angle of radiation of 20 degrees (for a dipole antenna 3/4 wavelength high) but, since LARAE's individual components show considerable directivity, this 20-degree angle is depressed somewhat to approximately 18 degrees above the horizon for this 75-meter application.[5] DX performance on 75 meters will generally be excellent with a wave angle of 18 degrees. See Fig. 1.

How does LARAE work? Let's first review how gain is obtained with a collinear array of directive antennas (such as Yagis). The gain in dB over a dipole at the same height, for well-separated Yagis lined up in a row (see Fig. 2), will approximate the antenna gain of one Yagi, with roughly 3 dB to be added each time the number of Yagis in the array is doubled.[6] To illustrate, consider a simple Yagi to have 7-dB gain; then two Yagis, suitably separated, will show a gain of 7 + 3 = 10 dB as a two-unit collinear array of Yagis. Another 3 dB can now be had by doubling the size of the two-yagi array. That is, we now place four Yagis in a line. Consequently, we have a 13-dB-gain array! Offhand, it would seem that such a four-unit system would act as an array in only two directions, since there are only two broadsides to a fixed straight line. A more careful examination of such an array will prove otherwise, and also reveal some very startling possibilities. (All gain figures are with reference to a dipole of the same height as the center height of the 3/4-wavelength Yagi array.)

Let's return to the simple case of two 7-dB Yagis, each at 3/4-wavelength above ground. Also,

let's make these Yagis into rotatable beam antennas. To start with, let's also place these two towers exactly one wavelength apart on a north-south line and point them 60 degrees to the east of north (toward Africa, from California).

The Geometry of LARAE

This arrangement of the two Yagis is shown in both elevation and plan view in Fig. 3A. Our next assumption is that we feed these two Yagis precisely 180 degrees out of phase; this is done easily by selecting proper lengths of feed lines from the final amplifier. We can readily visualize how far out from antenna B a wave will progress in exactly 1/2 cycle if part of a circle is drawn around the antenna (with circle radius equal to half a wavelength). It will travel a half wavelength through the air, as shown by the arrow BC in Fig. 3B.

Let me now draw your attention to the line drawn through antenna A and tangent to the circle around antenna B. We will call this line the wave-

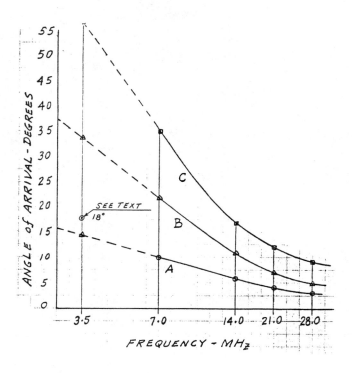

Fig. 1 — Vertical angle of arrival of sky-wave signals. Data below 7 MHz is extrapolated. Curve A, angle above which signals arrive 99% of the time. Curve B, angle above which signals arrive 50% of the time. Curve C, angle below which signals arrive 99% of the time.

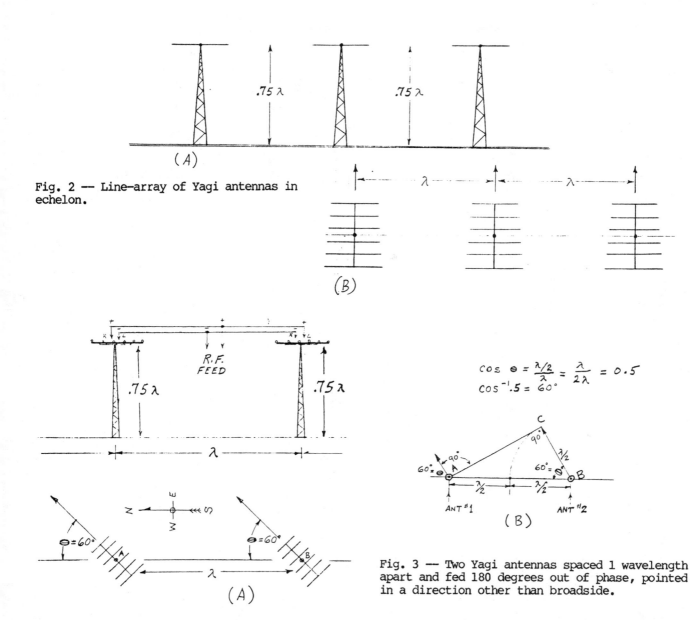

Fig. 2 — Line-array of Yagi antennas in echelon.

$$\cos \theta = \frac{\lambda/2}{\lambda} = \frac{\lambda}{2\lambda} = 0.5$$
$$\cos^{-1}.5 = 60°$$

Fig. 3 — Two Yagi antennas spaced 1 wavelength apart and fed 180 degrees out of phase, pointed in a direction other than broadside.

front line, since both Yagis are firing exactly perpendicular to this line AC. My next point is crucial, so follow me carefully. Were the two antennas to be fed in phase, then by the time the wave has traveled from antenna B to point C (1/2 cycle of time), the antenna at A is ready to reverse the current in its elements. The radiation field from the two antennas would now **cancel** each other in the forward direction of the two Yagis.

Fortunately, we can cure this ambiguity simply by transposing the feed wires to antenna B, as shown in Fig. 3A. This will assure that a wave arriving at point C (from antenna B) will have the same phase as the wave just starting to leave antenna A. We now have the two antennas working together to produce an additive, common wavefront described by line AC. This composite wave will now progress outward in a direction perpendicular to line AC.

Sometimes, knowing the answer beforehand helps to illustrate the solution to a geometry problem. This was the case when I chose 60 degrees as the common firing angle for our two antennas. Had we not known what it would turn out to be, we could have found out as follows:

$$\text{cosine } \theta = \frac{\lambda/2}{\frac{\lambda}{2} + \frac{\lambda}{2}} = \frac{\lambda/2}{\lambda} = \frac{\lambda}{2\lambda} = \frac{1}{2} = 0.50$$

(Eq. 1)

The angle whose cosine = 0.50 is 60 degrees, or $\cos^{-1} \theta$ = 60 degrees. What is the meaning of all of this geometry? Simply stated, we have now discovered that two high-performance rotary antennas can be positioned for complete array-effect additive gain for two separate angles from line AC, namely 60 degrees and 90 degrees. Also, when these two antennas are both set at zero degrees in azimuth, they will be additive if they are fed in phase. This gives us direction no. 3. Not bad, so far! (I will discuss this third case later on.)

By now you are asking yourself, "What if we line up three or even four Yagis on a single straight line, all one wavelength apart?" The answer is, "More gain at the same three angles as before, that is, 90, 60 and 0 degrees." For instance, four Yagis (each with 7.0 dBd gain) would show an approximate gain of 7.0 + 3.0 + 3.0 = 13 dB over a dipole. At 3.8 MHz, this would be fantastic!

The next question a DXer might ask is, "How can we get these separate rotaries to work together at angles besides 60 degrees?" (We will ignore 0 and 90 degrees for the time being). Well, let's try spacing our two Yagis out even more, to 1-1/2 wavelengths apart. See Fig. 4.

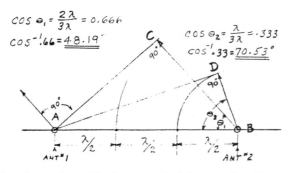

Fig. 4 — Two Yagis spaced 1.5 wavelengths apart. Two "angles of joy" exist. (See Text.)

Now our geometry has changed a bit, but it still looks familiar. We have two "angles of joy," θ1 and θ2. θ1 has a vector angle of less than 60 degrees, and θ2 seems somewhat greater than 60 degrees. Again, let's fall back on our high school geometry:

$$\text{cosine } \theta 1 = \frac{\lambda}{3/2 \lambda} = \frac{2\lambda}{3\lambda} = \frac{2}{3} = 0.667 \qquad \text{(Eq. 2)}$$

$$\theta 1 = \cos^{-1} 0.667 = 48.19 \text{ degrees}$$

$$\text{cosine } \theta 2 = \frac{\lambda/2}{3/2 \lambda} = \frac{\lambda}{3\lambda} = \frac{1}{3} = 0.333 \qquad \text{(Eq. 3)}$$

$$\theta 2 = \cos^{-1} 0.333 = 70.53 \text{ degrees}$$

Now let's examine Fig. 5A. A look at this drawing should definitely raise the blood pressure a bit and rev up the old ticker of a true DXer! As anticipated, we now have three angles of joy. Without delay, let's calculate them:

$$\text{cosine } \theta 1 = \frac{3 \times \frac{\lambda}{2}}{4 \times \frac{\lambda}{2}} = \frac{3\lambda}{4\lambda} = \frac{3}{4} = 0.75 \qquad \text{(Eq. 4)}$$

$$\theta 1 = \cos^{-1} 0.75 = 41.41 \text{ degrees}$$

$$\text{cosine } \theta 2 = \frac{2 \times \lambda/2}{4 \times \frac{\lambda}{2}} = \frac{2\lambda}{4\lambda} = \frac{2}{4} = 0.5 \qquad \text{(Eq. 5)}$$

$$\theta 2 = \cos^{-1} 0.5 = 60.00 \text{ degrees}$$

$$\text{cosine } \theta 3 = \frac{\lambda/2}{4 \times \frac{\lambda}{2}} = \frac{\lambda}{4\lambda} = \frac{1}{4} = 0.25 \qquad \text{(Eq. 6)}$$

$$\theta 3 = \cos^{-1} 0.25 = 75.52 \text{ degrees}$$

I expanded the spacing between Yagis, all the way out to three wavelengths. A general rule developed: Not counting 0 degrees or 90 degrees, the number of discrete, useful angles of joy are related to the spacing between antennas by the equation:

$$N = S - 1 \qquad \text{(Eq. 7)}$$

where S is the number of 1/2-wavelength spaces between antennas. (See Figs. 2 and 6.)

An extension of our geometry will produce Figs. 6 and 7, along with two more sets of angles.

For Fig. 6 we have: 36.87 degrees
53.13 degrees
66.42 degrees } S = 5
78.46 degrees

For Fig. 7 we have: 33.56 degrees
48.19 degrees
60.00 degrees } S = 6
70.53 degrees
80.41 degrees

Remember, S always equals the number of 1/2-wavelength spacings between antennas.

LARAE's Radiation Patterns

At about this point in LARAE's development, I began to wonder what the azimuthal pattern would look like. I had to know the half-power beamwidth, to see if the array would cover the required DX headings in a given azimuthal quadrant, such as 0 degrees to 90 degrees. Would this beamwidth change substantially as I went from one angle of joy to the next? Professor Simon Lheto, OH8OS, is a true DXer and one of the world's best antenna men. Simon agreed to program LARAE's array characteristics on the big computer at the University of Oulu as free time became available.

After a few months Simon sent me beautiful pattern printouts, accurate in every detail. The results were both good and bad. (See Figs. 8 and 9.) The broadside pattern was only 8 degrees wide at the half-power points, and there were two somewhat significant sidelobes at 22 degrees either side of the main lobe. This was for a LARAE of three 7-dBd Yagis at 3-wavelength antenna spacing. Our "cannon" had turned into a "laser-gun," with large gaps in the azimuth not covered! The good news was approximately 12.0 dB of gain over a dipole. Also, there were no significant lobes to the rear half of the pattern.

About this time Simon was able to visit me, and we discussed how to improve LARAE. We finally decided that we would increase the width of the main lobe by decreasing the spacing of LARAE. This meant a decrease in antenna spacing from 3 toward 2 wavelengths. The final spacing chosen was 2.103 wavelengths. (I'll explain this odd value later.) Theory also indicated that the sidelobes at plus and minus 22 degrees would be attenuated, as a further benefit. Simon returned to the computer in northern Finland, and I returned to the drawing board.

Our last few hours together resulted in our agreeing to incorporate delay-line phasing into the LARAE. The main idea was to wobble each of the above calculated angles (vector directions) of Fig. 5A a couple of steps, both minus then plus, as we

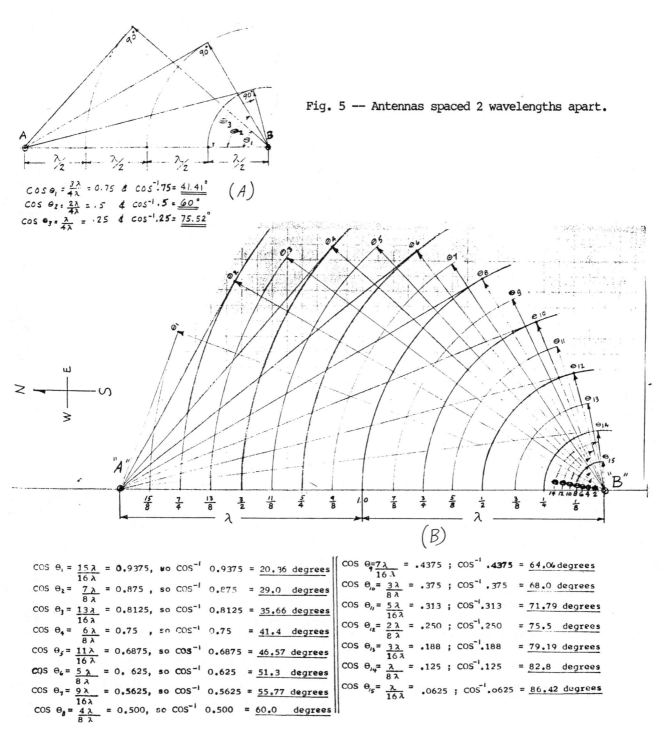

Fig. 5 — Antennas spaced 2 wavelengths apart.

$$\cos \theta_1 = \frac{3\lambda}{4\lambda} = 0.75 \ \& \ \cos^{-1}.75 = \underline{41.41^\circ}$$
$$\cos \theta_2 = \frac{2\lambda}{4\lambda} = .5 \ \& \ \cos^{-1}.5 = \underline{60^\circ}$$
$$\cos \theta_3 = \frac{\lambda}{4\lambda} = .25 \ \& \ \cos^{-1}.25 = \underline{75.52^\circ}$$

(A)

(B)

$\cos \theta_1 = \frac{15\lambda}{16\lambda} = 0.9375$, so $\cos^{-1} 0.9375 = \underline{20.36 \text{ degrees}}$

$\cos \theta_2 = \frac{7\lambda}{8\lambda} = 0.875$, so $\cos^{-1} 0.875 = \underline{29.0 \text{ degrees}}$

$\cos \theta_3 = \frac{13\lambda}{16\lambda} = 0.8125$, so $\cos^{-1} 0.8125 = \underline{35.66 \text{ degrees}}$

$\cos \theta_4 = \frac{6\lambda}{8\lambda} = 0.75$, so $\cos^{-1} 0.75 = \underline{41.4 \text{ degrees}}$

$\cos \theta_5 = \frac{11\lambda}{16\lambda} = 0.6875$, so $\cos^{-1} 0.6875 = \underline{46.57 \text{ degrees}}$

$\cos \theta_6 = \frac{5\lambda}{8\lambda} = 0.625$, so $\cos^{-1} 0.625 = \underline{51.3 \text{ degrees}}$

$\cos \theta_7 = \frac{9\lambda}{16\lambda} = 0.5625$, so $\cos^{-1} 0.5625 = \underline{55.77 \text{ degrees}}$

$\cos \theta_8 = \frac{4\lambda}{8\lambda} = 0.500$, so $\cos^{-1} 0.500 = \underline{60.0 \text{ degrees}}$

$\cos \theta_9 = \frac{7\lambda}{16\lambda} = .4375$; $\cos^{-1} .4375 = \underline{64.06 \text{ degrees}}$

$\cos \theta_{10} = \frac{3\lambda}{8\lambda} = .375$; $\cos^{-1} .375 = \underline{68.0 \text{ degrees}}$

$\cos \theta_{11} = \frac{5\lambda}{16\lambda} = .313$; $\cos^{-1} .313 = \underline{71.79 \text{ degrees}}$

$\cos \theta_{12} = \frac{2\lambda}{8\lambda} = .250$; $\cos^{-1} .250 = \underline{75.5 \text{ degrees}}$

$\cos \theta_{13} = \frac{3\lambda}{16\lambda} = .188$; $\cos^{-1} .188 = \underline{79.19 \text{ degrees}}$

$\cos \theta_{14} = \frac{\lambda}{8\lambda} = .125$; $\cos^{-1} .125 = \underline{82.8 \text{ degrees}}$

$\cos \theta_{15} = \frac{\lambda}{16\lambda} = .0625$; $\cos^{-1} .0625 = \underline{86.42 \text{ degrees}}$

progress from north to east. We do electrically what we cannot do mechanically at the closer 2-wavelength spacing. Namely, we now have in-phase radiation on the 1/8-wavelength circles, where before we had that only at the 1/2-wavelength circles. (See Fig. 5B). This delay-line phasing gives us a greater number of discrete lobe angles which will be more uniformly spaced throughout a 90-degree sector (16 angles, not counting north).

Steering by Turning and Phasing — With Optimum Gain

The three-antenna LARAE, by its natural symmetry, simplified our phasing study somewhat. At this point it is imperative to fix in your mind that delays and advances of phase will be applied at the end Yagis only. The center Yagi is the pivot. This will have the effect of tilting the antenna line (tower line) plus or minus a few degrees about the center tower as a pivot. In other words, it acts as if the end towers were shifted from the antenna line, one above and the other below. To further establish this important concept, imagine the tower line as running east and west. Broadside, then, for the main lobe would be dead on the geographic north pole. A 10-degree counter-clockwise rotation of the line would cause the broadside lobe to fire on Bombay, India, while a ten-degree rotation clockwise would target northern Europe. Our next task was to deal with this complex phasing problem.

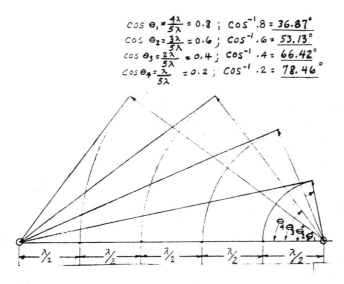

$$\cos \theta_1 = \frac{4\lambda}{5\lambda} = 0.8 \; ; \; \cos^{-1}.8 = \underline{36.87°}$$
$$\cos \theta_2 = \frac{3\lambda}{5\lambda} = 0.6 \; ; \; \cos^{-1}.6 = \underline{53.13°}$$
$$\cos \theta_3 = \frac{2\lambda}{5\lambda} = 0.4 \; ; \; \cos^{-1}.4 = \underline{66.42°}$$
$$\cos \theta_4 = \frac{\lambda}{5\lambda} = 0.2 \; ; \; \cos^{-1}.2 = \underline{78.46°}$$

Fig. 6 — Antennas spaced 2.5 wavelengths apart.

Graph of Phase Angle of Weights

The graph in Fig. 10A was prepared by Simon's computer crew. This graph is the result of an X-, Y- and Z- coordinate plot of the advancing, unified wave front when, and only when, the three antenna's phases have been adjusted to provide maximum field intensity at a far-away place in a given direction. Weight, as used here, defines the combination of phase and power at each antenna. The power part is simple; we chose to put 1/3 the RF power into each antenna. We also chose to vary the phase in increments of 45 degrees to step from one angle of joy to the next.

Now is the time that the reader fix in his or her mind how LARAE is steered. First, the phase angle on the end antennas is advanced on one and retarded on the other by 22-1/2 degrees. This gives a total change, between the two end antennas, of 45 degrees (1/8 wavelength). Once this phasing change has been made, then all three antennas are rotated to a common angle where the distant-point field strength is maximized. This specific angle can be computed from an expanded version of Fig. 5A, as shown in Fig. 5B. Basically, we insert new partial circles about point B showing wave progressions in increments of 1/8 wavelength. We can then solve arccos x for every total change of 1/8 wavelength. Since we have the antennas spaced by 2 wavelengths, we have 16 angles that we can solve for as separate

angles of joy. This, of course, totals to sixteen firing angles for a 90-degree quadrant on a great-circle map.

The complete success of any project depends on close attention to important details. One such detail has to do with how LARAE will work when all the Yagis are firing precisely down the line of the towers. In this case the phasing should be zero degrees (firing line = tower line). Fig. 11 shows a side view of the three towers and depicts the take-off angle of main-lobe radiation at 18 degrees from each of the three antennas, all being fed in phase. Simple geometry shows that the three towers must be spaced 2.103 wavelengths apart in order for the advancing wave front from each antenna to combine precisely with those from the other two and produce an overall wave front with a take-off angle of 18 degrees above the earth.

At this unique angle, LARAE is working as a stacked array, for all practical purposes. The vertical distance between the energy beams is 0.65 wavelength, as shown in Fig. 11. As LARAE turns away from this unique zero-degree angle, it gradually goes from a stacked array to a collinear array. If you ponder this awhile, you will conclude that the graph of phase angle of weights (Fig. 10A) must be solved as a problem in three-dimensional geometry in order that each firing angle be precisely computed. Using the plane of the earth's surface only would introduce serious errors for all angles near θ = zero. Simon's crew **did** use the solid geometry! (The actual math and solid geometry used to compute these sixteen solid angles are not shown in this article.) The horizontal components of each of these solid angles are shown on the graph in Fig. 10A as the abscissa, while phase-angle differences between end antennas are shown as ordinates. Intercepts of the five horizontal lines on the graph delineate the sixteen firing angles (optimum angles) for the array, and also show the required phase difference between the end antennas for each angle.

Engineering Implementation

The towers used at W6KPC are the 180-ft Tri-Ex "Clementower," which are rotated from the bottom with computer-controlled stepping motors. The amount (angle) of rotation is controlled by an Intel 8085-2 central processing unit, which is part of an Intel ICS 580 main-frame computer equipped with ROM, RAM, and analog-to-digital converters (and vice-versa) as well as clocks, printers, keyboard, video monitor, and so on. This industrial control computer is programmed to (1) rotate all Yagis, in unison, to the

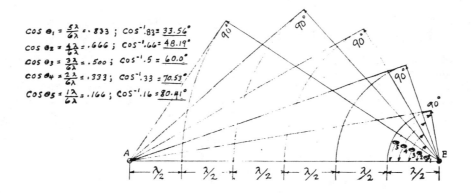

$$\cos \theta_1 = \frac{5\lambda}{6\lambda} = .833 \; ; \; \cos^{-1}.83 = \underline{33.56°}$$
$$\cos \theta_2 = \frac{4\lambda}{6\lambda} = .666 \; ; \; \cos^{-1}.66 = \underline{48.19°}$$
$$\cos \theta_3 = \frac{3\lambda}{6\lambda} = .500 \; ; \; \cos^{-1}.5 = \underline{60.0°}$$
$$\cos \theta_4 = \frac{2\lambda}{6\lambda} = .333 \; ; \; \cos^{-1}.33 = \underline{70.53°}$$
$$\cos \theta_5 = \frac{1\lambda}{6\lambda} = .166 \; ; \; \cos^{-1}.16 = \underline{80.41°}$$

Fig. 7 — Antennas spaced 3 wavelengths apart.

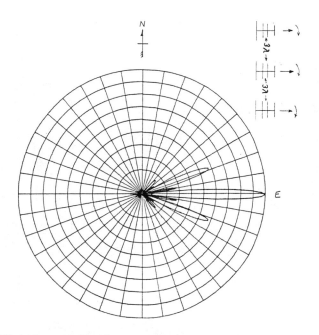

Fig. 8 — Radiation pattern of 3-wavelength, 3-Yagi LARAE firing broadside.

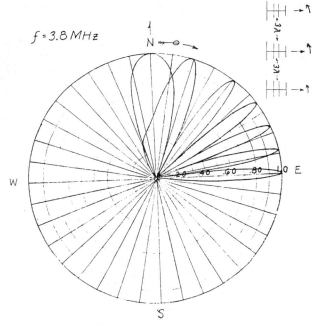

Fig. 9 — Radiation patterns 3-wavelength, 3-Yagi LARAE. Only major lobes are shown in one quadrant for different amounts of beam steering and Yagi rotation.

Consideration is being given to a voice-recognition command unit, coupled to a foot switch, which would instruct the computer to change the LARAE heading on voice command. Thus, when the foot switch is closed and the word "Finland" is spoken, the whole array would obediently rifle in on a bearing of 20 degrees from central California).

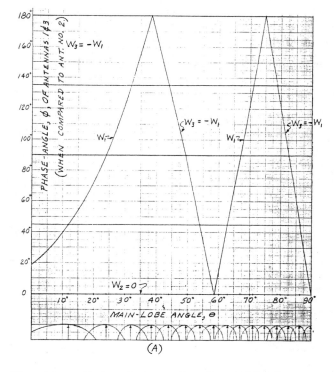

Fig. 10 — At A, phase-angle of weights for an antenna spacing of 2.1 wavelengths. At B, radiation patterns.

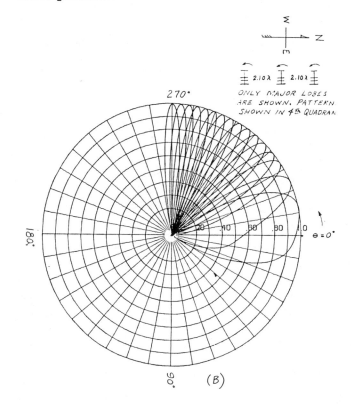

preprogrammed DX angle heading (as entered on the keyboard in real time or as called for by the clock); (2) set all relays on all feed systems to give the proper weights (phases) to the feed lines of the two end Yagis for the heading indicated; (3) manage all interlocks to prevent hot switching of RF relays; and (4) measure the signal strength (frequently) while in the receive mode, and then instantly (and automatically) wobble the main lobe to see if signals are perhaps arriving on a skewed path. If skewed path is detected, the computer will then rotate all Yagis to the correct heading for maximum incoming signal strength.

Antennas and Feeds

The LARAE at W6KPC was designed for 3.8 MHZ as a center frequency. The three Yagi antennas are KLM four-element Yagis constructed on a 76-foot boom with a gain of 7.25 dB over a dipole. The drive-point impedance is nonreactive at 200 ohms, using a two-element log-periodic cell as the double driven dipole. Each Yagi antenna is equipped with a 4:1 balun to step the antenna's balanced 200-ohm feed impedance point down to 50 ohms unbalanced. This allows the use of 50-ohm coaxial cable between the three towers and the transmitter.

Fig. 12 shows how the three antennas are fed as an array. Note that the two end antennas are fed through the two relay boxes, while the center antenna is fed directly as the pivot antenna. The same system that was developed to feed the collinear Yagi Sextet's three separate bays (levels)\7 is utilized here to feed LARAE's three separate antennas. Notice the manner in which K6SSJ and W6KPC devised the quarter-wave matching sections so the transmitter constantly looks at a 50-ohm feed-line impedance.\8

Table 1 is a tabulation of phase-angle weights and relay settings. Refer to Fig. 13 to see how the relay activations shown in the table insert the various line sections, causing the desired delays in the end antennas, no. 1 and no. 3. Notice, particularly, that line sections are used two at a time to give a total phasing step of 45 degrees, each of the two sections representing 22-1/2 degrees.

LARAE Radiation Pattern and Gain

The KLM four-element Yagi used as the unit antenna in LARAE has somewhat shortened (folded) elements approximately 90 feet long, and a boom length of 76 feet. It uses a broadbanded feed system made famous by KLM, the double driven dipole, or two-element log cell. KLM specifies a power gain of 7.25 dB over a dipole for this antenna. The horizontal-plane pattern for this antenna is shown in Fig. 14. I developed an enormous respect for this Yagi three winters ago. I listened all winter long to Arnold, W2HCW, work the world with one, receiving incredible reports. W2HCW dominated the 75-meter band from the U. S. A. at that time.

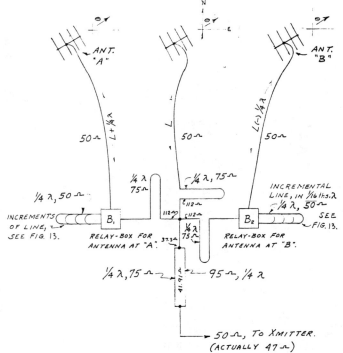

Fig. 12 — Feed arrangement of a 3-Yagi LARAE.

Fig. 8 shows the typical interferometer grating pattern of three widely spaced Yagi antennas when used in a broadside array.\9 This is the broadside pattern (θ = 90 degrees) for the antenna diagrammed in Fig. 7. Fig. 9 shows the main lobes of θ1 through θ5 as calculated from Fig. 7. Also, there is a fat north lobe and a skinny east lobe for the stacked (north) and broadside special conditions. This completes the family of seven lobes shown in Fig. 9. Remember, however, that the five angles cal-

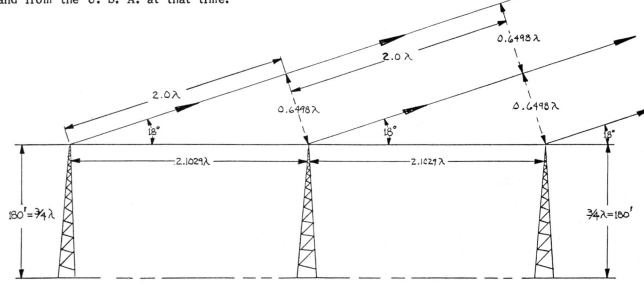

Fig. 11 — LARAE firing down the tower line. This is the "stacked" mode of operation.

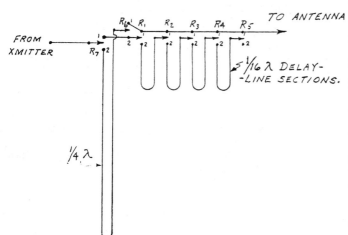

Fig. 13 — Diagram of relay box for a 3-Yagi LARAE. Two boxes are used, one at each end Yagi.

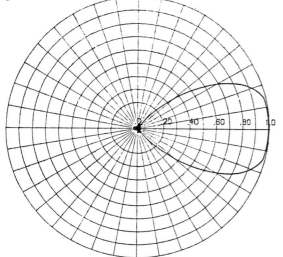

Fig. 14 — Approximate horizontal radiation pattern of a 3.8-MHz Yagi (single antenna). Performance data is summarized below.

Manufacturer: KLM
Measured gain: 7.25 dBd
Half-power beamwidth: 70°
Front-to-back ratio: 25 dB
Front-to-side ratio: Greater than 30 dB

culated in Fig. 7 are not the result of three-dimensional solid geometry, but of only plane geometry, for the sake of simplifying the theory of how we get natural phase coincidence at certain angles. When Simon ran these computations on the computer we came up with considerable variance in these angles when solid geometry was applied.

Summary

Sixteen angles, it turns out, give excellent pattern overlapping such that there is never a drop in far-point field strength versus beam heading of more than 1 dB. This is shown by how the patterns intersect each other at the bottom of Fig. 10A. As a net result, we have a practical antenna array that will show a real-world gain of approximately 12 dB over a dipole at the same antenna height. Also, the

antenna array is steerable over 360 degrees of azimuth in a step-wise manner that approaches the "all azimuths available" advantage of a single Yagi rotary-beam antenna.

Fig. 1 shows that obtaining a radiation angle below 15 degrees on 75 meters is not cost effective, since signals arrive above 14.5 degrees for 99% of the time.\10 A good height for a horizontally polarized 75-meter LARAE would appear to be 0.75 wavelength, which is 180 feet at 3.8 MHz. This height will give a vertical radiation angle of 18 degrees.\11,12 This 18-degree angle will come as no surprise to veteran DXers who already know that one does not need as low a radiation angle on 75 as is necessary on 10, 15, and 20 meters. I want to emphasize that the points on the graph of Fig. 1 that fall on the 3.5 MHz ordinate are extrapolated, as shown by the dotted lines. There is little reason, however, to challenge the validity of these 3.5-MHz points.

Earlier in this article reference was made to the fact that the LARAE would reduce both QRM and QRN. Some contest operators will argue for a broad azimuth pattern accompanied by a narrow vertical pattern.\13 This may be okay on the 20- to 10-meter bands, but is patently undesirable on 75 meters. The reasons are simple: (1) Static crashes are much worse on 75 meters than on the higher frequency bands, and this kind of QRN comes from many directions. (2) Nonamateur QRM and QRN may come from any direction. (3) Omnidirectional amateur QRM is often fierce in the relatively narrow DX portion of 75 meters. (4) High-angle radiation should not be suppressed unduly since we have previously noted the likelihood of high-angle arrival of DX signals on 3.5 MHz. (This 3/4-wavelength-high array does have a high-angle, vestigal, minor lobe, which is useful for close-in DX QSO's).

Epilog

As this article was being readied for mailing to ARRL Headquarters, Simon, OH8OS and I decided to make one more computer run. This run was for a Yagi antenna spacing of only 1.05 wavelengths and using three antennas, as before. The results were wonderful! The reduction in array gain is negligible, the minor lobes all but disappear, and each directional lobe is fattened to the extent that eight beams fill a quadrant with excellent overlap. This is all illustrated in Figs. 15 (which shows the geometrical math) and 16. Fig. 16A shows a polar plot of the angles calculated and shown by Fig. 15. Curves are added to this plot for 0 and 90 degrees, plus the first lobe in each neighboring quadrant, to illustrate symmetry for a whole circle. Fig. 16B is included to show how minimal the worst case really is (90 degree, broadside firing), as far as interferometer-type grating lobes are concerned. Fig. 17 shows a graph of phase angle of weights for 1.05 wavelengths.

Acknowledgements

The basic idea for the LARAE occurred to me in the early 1970s, when I made a few sketches. Calculations did not start until about 1975, when I tried the idea with a couple of professional antenna-array engineers. These engineers were intrigued with the LARAE concept and urged me to pursue the idea further. When Simon, OH8OS, visited me in the mid-70s, I showed him the drawings, charts and calculations. He became enthused and volunteered to run some computer calculations and pattern plots. Simon has been a great help to me in properly sorting out some of the finer details and concepts that were involved in understanding the LARAE. I wish to thank one of Simon's associates, Rauno Suikola, for his excellent work on the computer programming associated with this project.

References

\1 Clement, A. J. F., "The Yagi Dagi", **QST,** September 1951.

\2 Clement, A. J. F., "The Driven Beast" **QST,** May 1958.

\3 Clement, A. J. F., "The Collinear Yagi Quartet" **QST,** November 1969.

\4 Clement, A. J. F., "The Collinear Yagi Sextet" **QST,** June 1980.

\5 Bachelor, W. B., "Combined Vertical Directivity" **QST,** February 1981.

\6 The **The ARRL Antenna Book,** 13th Ed., p. 135, Fig. 4-9.

\7 See Ref. 4.

\8 Kraus, John D., **Antennas** (McGraw-Hill), p. 437.

\9 Jasik, H., **Antenna Engineering Handbook** (McGraw-Hill), p. 29-20.

\10 See Ref. 6, pages 18-19 and Table 1-1.

\11 See Ref. 6, p. 49, Fig. 2-42.

\12 See Ref. 5.

\13 Lawson, James L., "Yagi Antenna Design," **Ham Radio** p. 23, November 1980.

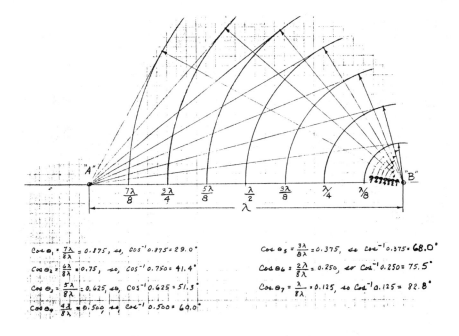

Fig. 15 — LARAE with 1.05-wavelength spacing and phasing such that radio energy arriving at antenna A lags energy arriving at antenna B by multiples of 1/8 to 7/8 wavelength.

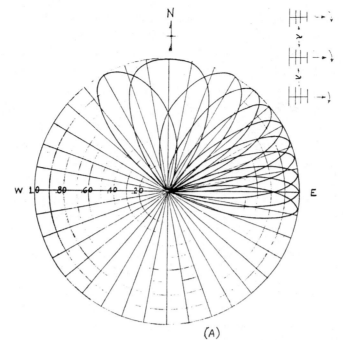

(A)

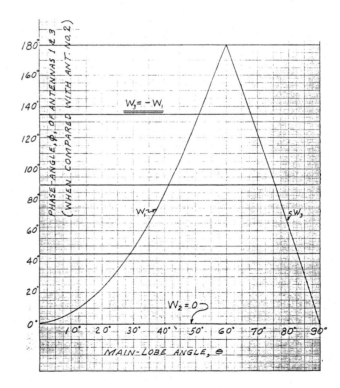

18° 18° at ½ power points.

θ = 90° (BROADSIDE FIRING)

(B)

Fig. 16 — Radiation patterns of LARAE with 1.05-wavelength spacing and increments of 1/8-wavelength phasing. Height is 3/4 wavelength above ground.

Fig. 17 — Phase-angle of weights for an antenna spacing of 1.05 wavelengths.

TABLE 1

TABLE OF PHASE-ANGLES AND RELAY SETTINGS
(Antennas Spaced At 2.10 λ)

Horiz. Component Azimuth - Angle, θ	Ant. "A" Phase, (Degrees)	Ant. "B" Phase, (Degrees)	Relays - Settings, Ant. "A" 1 2 3 4 5 6 7	Relays - Settings, Ant. "B" 1 2 3 4 5 6 7	Total (A + B) Phase
θ16 (East) = 90.00°	-0-	-0-	1 - - - - - 1	2 2 2 2 1 2 2	-0-
θ15 = 86.42°	- 22.5	+ 22.5	2 1 - - - - 1	2 2 2 1 - 2 2	45°
θ14 = 82.80°	- 45.0	+ 45.0	2 2 1 - - - 1	2 2 1 - - 2 2	90°
θ13 = 79.19°	- 67.5	+ 67.5	2 2 2 1 - - 1	2 1 - - - 2 2	135°
θ12 = 75.50°	- 90.0	+ 90.0	1 - - - - 2 2	2 2 2 2 1 - 1	180°
θ11 = 71.79°	-112.5	+112.0	2 1 - - - 2 2	2 2 2 1 - - 1	225°
θ10 = 68.00°	-135.0	+135.0	2 2 1 - - 2 2	2 2 1 - - - 1	270°
θ 9 = 64.06°	-157.5	+157.5	2 2 2 1 - 2 2	2 1 - - - - 1	315°
θ 8 = 60.00°	180.0	180.0	2 2 2 2 1 2 2	1 - - - - - 1	360°
θ 7 = 55.77°	+ 22.5	- 22.5	2 2 2 1 - 2 2	2 1 - - - - 1	405°
θ 6 = 51.30°	+ 45.0	- 45.0	2 2 1 - - 2 2	2 2 1 - - - 1	450°
θ 5 = 46.57°	+ 67.5	- 67.5	2 1 - - - 2 2	2 2 2 1 - - 1	495°
θ 4 = 41.40°	+ 90.0	- 90.0	2 2 2 2 1 - 1	1 - - - - 2 2	540°
θ 3 = 35.66°	+112.5	-112.5	2 2 2 1 - - 1	2 1 - - - 2 2	585°
θ 2 = 29.00	+135.0	-135.0	2 2 1 - - - 1	2 2 1 - - 2 2	630°
θ 1 = 20.36°	+157.5	-157.5	2 1 - - - - 1	2 2 2 1 - 2 2	675°
θ 0 (North) = -0-	180.0	180.0	1 - - - - - 1	2 2 2 2 1 2 2	720°

NOTE: Transmission line to antenna "A" is 1/4 λ longer then to middle antenna. Also, line to antenna "B" is 1/4 λ shorter than to middle antenna.

Multiband Antennas

A Great 10 Through 40 Portable Antenna

By Edward L. Henry,* KØGPD

*25 Gold Run Drive, St. Peters, MO 63376

For years I have seen local clubs construct Field Day antennas. Two or more vehicles were usually necessary to transport the array of antenna equipment to the site. Then a group of people would labor for several hours to erect the antennas. It can be a big, tiring job. Obviously, this is not the way to do it. In a real emergency, you do not want to be delayed. You need the antenna now!

I designed a light-weight 10- through 40-meter inverted-V antenna that can be transported easily in the trunk of a small car. The elements are full-length dipoles (no traps) and can be erected by one man and ready to operate in 15 minutes. You do not need trees or existing structures to set it up, and it handles full legal power.

The antenna performs so well that I use it for other occasions such as demonstrating Amateur Radio to schools and scout meetings. A park on a Sunday afternoon is a pleasant place to set up and talk to the world. Depending on where you are, spectators will be attracted and it is a good way to promote our hobby. It can even be set up on concrete parking lots, where nails can be driven into the cracks for anchors. An antenna tuner is not necessary, but using one will give you a perfect match.

All construction information is given on the drawings. Most materials are inexpensive and available at hardware stores. The antenna pole consists of six standard Radio Shack 1-1/4 inch diameter TV mast sections. The mast sections are 4 to 5 feet long and are made of aluminum or steel. Aluminum masts are preferred because of their light weight. Insulated wire is used for the antenna elements in the event one element touches another when the wind blows.

I have experimented with many batteries (lead-acid and NiCd) for emergency power. The marine type battery used by motorboaters outperforms them all. When you use a 100- to 200-W mobile type transceiver, a 12-V marine battery can operate 6 or more hours with 50% talk time. A 3-A charger is great for recharging the battery. The charger and a specific gravity tester can be taken on vacations. Remember to keep sparks away from a charging battery! (Using a newspaper, always fan the air away from the battery and have equipment turned off before making any electrical connections to or near the battery.)

No matter how many antennas you own, this portable multiband antenna will be a welcome

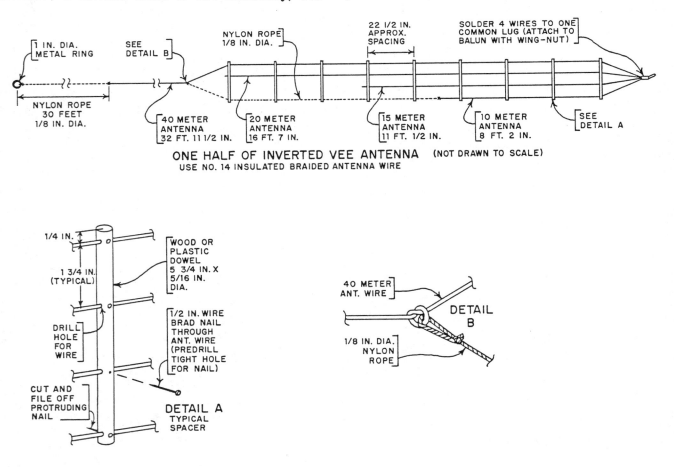

ONE HALF OF INVERTED VEE ANTENNA (NOT DRAWN TO SCALE)
USE NO. 14 INSULATED BRAIDED ANTENNA WIRE

addition. Except for the pole sections, the antenna and hardware fit into a one-cubic-foot cardboard box. Keep it stored in the corner of your garage and it will be ready to go on a moment's notice.

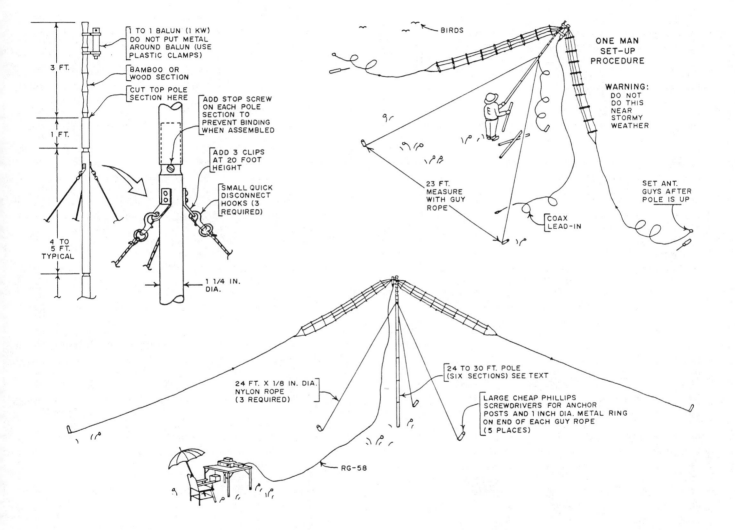

1 TO 1 BALUN (1 KW)
DO NOT PUT METAL
AROUND BALUN (USE
PLASTIC CLAMPS)

BAMBOO OR
WOOD SECTION

CUT TOP POLE
SECTION HERE

ADD STOP SCREW
ON EACH POLE
SECTION TO
PREVENT BINDING
WHEN ASSEMBLED

ADD 3 CLIPS
AT 20 FOOT
HEIGHT

SMALL QUICK
DISCONNECT
HOOKS (3
REQUIRED)

3 FT.

1 FT.

4 TO
5 FT.
TYPICAL

1 1/4 IN.
DIA.

BIRDS

ONE MAN
SET-UP
PROCEDURE

WARNING:
DO NOT
DO THIS
NEAR
STORMY
WEATHER

23 FT.
MEASURE
WITH GUY
ROPE

COAX
LEAD-IN

SET ANT.
GUYS AFTER
POLE IS UP

24 FT. X 1/8 IN. DIA.
NYLON ROPE
(3 REQUIRED)

24 TO 30 FT. POLE
(SIX SECTIONS) SEE TEXT

LARGE CHEAP PHILLIPS
SCREWDRIVERS FOR ANCHOR
POSTS AND 1 INCH DIA. METAL RING
ON END OF EACH GUY ROPE
(5 PLACES)

RG-58

The G5RV Multiband Antenna . . . Up-to-Date

By Louis Varney,* G5RV

*82 Folders Lane, Burgess Hill, W. Sussex RH15 ØDX, United Kingdom

Adapted from an article of the same title in **Radio Communication,** July 1984, pp. 572-575.

The G5RV antenna, with its special feeder arrangement, is a multiband center-fed antenna capable of efficient operation on all HF bands from 3.5 to 28 MHz. Its dimensions are specifically designed so it can be installed in areas of limited space, but which can accommodate a reasonably straight run of about 102 ft for the flat-top. Because the most useful radiation from a horizontal or inverted-V resonant antenna takes place from the center two-thirds of its total length, up to one-sixth of this total length at each end of the antenna may be dropped vertically, semi-vertically, or bent at a convenient angle to the main body of the antenna without significant loss of effective radiation efficiency. For installation in very limited areas, the dimensions of both the flat-top and the matching section can be divided by a factor of two to form the half-size G5RV, which is an efficient antenna from 7 to 28 MHz. The full-size G5RV will also function on the 1.8-MHz band if the station end of the feeder (either balanced or coaxial type) is strapped and fed by a suitable matching network using a good earth connection or a counterpoise wire. Similarly, the half-size version may be used on the 3.5- and 1.8-MHz bands.

In contradistinction to multiband antennas in general, the full-size G5RV antenna was **not** designed as a λ/2 dipole on the lowest frequency of operation, but as a 3λ/2 center-fed long-wire antenna on 14 MHz, where the 34 ft open-wire matching section functions as a 1:1 impedance transformer. This enables the 75-ohm twin lead, or 50/80-ohm coaxial cable feeder, to see a close impedance match on that band with a consequently low SWR on the feeder. However, on all the other HF bands, the function of this section is to act as a "make-up" section to accommodate that part of the standing wave (current and voltage components) which, on certain operating frequencies, cannot be completely accommodated on the flat-top (or inverted-V) radiating portion. The design center frequency of the full-size version is 14.150 MHz, and the dimension of 102 ft is derived from the formula for long-wire antennas which is:

$$\text{Length (ft)} = \frac{492(n - 0.05)}{f_{MHz}}$$

$$= \frac{492 \times 2.95}{14.15}$$

$$= 102.57 \text{ ft (31.27 m)}$$

where n = the number of half wavelengths of the wire (flat-top)

Because the whole system will be brought to resonance by the use of a matching network in practice, the antenna is cut to 102 ft.

As the antenna does not make use of traps or ferrite beads, the dipole portion becomes progressively longer in electrical length with increasing frequency. This effect confers certain advantages over a trap or ferrite-bead loaded dipole because, with increasing electrical length, the major lobes of the vertical component of the polar diagram tend to be lowered as the operating frequency is increased. Thus, from 14 MHz up, most of the energy radiated in the vertical plane is at angles suitable for working DX. Furthermore, the polar diagram changes with increasing frequency from a typical λ/2 dipole pattern at 3.5 MHz and a two λ/2 in-phase pattern at 7 and 10 MHz to that of a long-wire antenna at 14, 18, 21, 24 and 28 MHz.

Although the impedance match for 75-ohm twin lead or 80-ohm coaxial cable at the base of the matching section is good on 14 MHz, and even the use of 50-ohm coaxial cable results in only about a 1.8:1 SWR on this band, the use of a suitable matching network is necessary on all the other HF bands. This is because the antenna plus the matching section will present a **reactive** load to the feeder on those bands. Thus, the use of the correct type of matching network is essential in order to ensure the maximum transfer of power to the antenna from a typical transceiver having a 50-ohm coaxial (unbalanced) output. This means unbalanced input to balanced output if twin-wire feeder is used, or unbalanced to unbalanced if coaxial feeder is used. A matching network is also employed to satisfy the stringent load conditions demanded by such modern equipment that has an automatic level control system. The system senses the SWR condition present at the solid state transmitter output stage to protect it from damage, which could be caused by a reactive load having an SWR of more than about 2:1.[1]

The above reasoning does not apply to the use of the full-size G5RV antenna on 1.8 MHz, or to the use of the half-size version on 3.5 and 1.8 MHz. In these cases, the station end of the feeder conductors should be "strapped" and the system tuned to resonance by a suitable series-connected inductance and capacitance circuit connected to a good earth or counterpoise wire. Alternatively, an unbalanced-to-unbalanced type of matching network such as a T or L matching circuit can be used.[2] Under these conditions the flat-top (or inverted-V) portion of the antenna, plus the matching section and feeder, function as a Marconi or T antenna, with most of the effective radiation taking place from the vertical, or near vertical, portion of the system; the flat-top acts as a top-capacitance loading element. However, with the system fed as described above, very effective radiation on these two bands is obtainable even when the flat-top is as low as 25 ft above ground.

Theory of Operation

The general theory of operation has been explained above. The detailed theory of operation on each band from 3.5 to 28 MHz follows, aided by figures showing the current standing wave conditions on the flat-top, and the matching (or make-up) section. The relevant theoretical horizontal plane polar diagrams for each band may be found in any of the specialized antenna handbooks. However, it must be borne in mind that: (a) the polar diagrams generally shown in two dimensional form are, in fact, three dimensional (i.e., solid) figures around the plane of the antenna; and (b) all theoretical polar diagrams are modified by reflection and absorption effects of nearby conducting objects such as wire fences, metal house guttering, electric wiring systems, and even large trees. Also, the local earth conductivity will materially affect the actual polar radiation pattern produced by an antenna. Theoretical polar diagrams are based on the assumptions that an antenna is supported in "free space" above a perfectly conducting ground. Such conditions are obviously impossible of attainment in the case of typical amateur installations. What this means in practice is that the reader should not be surprised if **any** particular antenna in a typical amateur location produces contacts in directions where a null is indicated in the theoretical polar diagram, and perhaps not such effective radiation in the directions of the major lobes as theory would indicate.

3.5 MHz: On this band each half of the flat-top, plus about 17 ft of each leg of the matching section, forms a foreshortened or slightly folded up $\lambda/2$ dipole. The remainder of the matching section acts as an unwanted, but unavoidable reactance between the **electrical** center of the dipole and the feeder to the matching network. The polar diagram is effectively that of a $\lambda/2$ antenna. See Fig. 1.

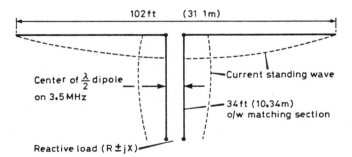

Fig. 1 — Current standing-wave distribution on the G5RV antenna and matching section at 3.5 MHz. The antenna functions as a $\lambda/2$ dipole partially folded up at the center.

7 MHz: The flat-top, plus 16 ft of the matching section, now functions as a partially folded up two half waves in phase antenna producing a polar diagram with a somewhat sharper lobe pattern than a $\lambda/2$ dipole because of its collinear characteristics. Again, the matching to a 75-ohm twin-lead or 50/80-ohm coaxial feeder at the base of the matching section is degraded somewhat by the unwanted reactance of the lower half of the matching section, but,

despite this, by using a suitable matching network, the system loads well and radiates very effectively on this band. See Fig. 2.

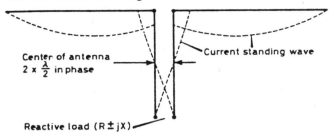

Fig. 2 — Current distribution on the antenna and matching section at 7 MHz. The antenna now functions as a collinear array with two half waves fed in phase.

10 MHz: On this band the antenna functions as a two half-wave in-phase collinear array, producing a polar diagram virtually the same as on 7 MHz. A reactive load is presented to the feeder at the base of the matching section but, as for 7 MHz, the performance is very effective. See Fig. 3.

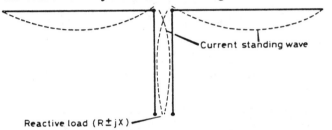

Fig. 3 — Current standing-wave distribution on the antenna and matching section at 10 MHz. The antenna functions as a collinear array with two half waves fed in phase.

14 MHz: At this frequency the conditions are ideal. The flat-top forms a $3\lambda/2$ long center-fed antenna which produces a multilobe polar diagram with most of its radiated energy in the vertical plane at an angle of about 14 degrees, which is effective for working DX. Since the radiation resistance at the center of a $3\lambda/2$ long-wire antenna supported at a height of $\lambda/2$ above ground of average conductivity is about 90 ohms, and the 34-ft matching section now functions as a 1:1 impedance transformer, a feeder of anything between 75 and 80 ohms characteristic impedance will see a nonreactive (i.e., resistive) load of about this value at the base of the matching section, so that the SWR on the feeder will be near 1:1. Even the use of 50-ohm coaxial feeder will result in an SWR of only about 1.8:1. It is assumed here that 34 ft is a reasonable average antenna height in amateur installations. See Fig. 4.

18 MHz: The antenna functions as two full-wave antennas fed in phase; it combines the broadside gain of a two-element collinear array with a somewhat lower zenithal angle radiation than a $\lambda/2$ dipole because of its long-wire characteristic. See Fig. 5.

21 MHz: On this band the antenna works as a long wire of five halfwaves, producing a multilobe polar diagram with effective low zenithal angle radiation. Although a high resistive load is presented to the feeder at the base of the make-up

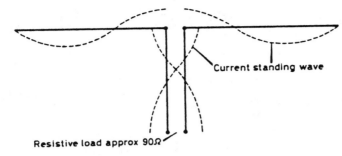

Current standing wave

Resistive load approx 90 Ω

Fig. 4 — Current standing-wave distribution on the antenna and matching section at 14 MHz. In this case the antenna functions as a center-fed long wire of three half waves out of phase. The matching section now functions as a 1:1 impedance transformer, presenting a resistive load of approximately 90 ohms at the lower end.

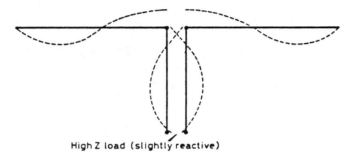

High Z load (slightly reactive)

Fig. 5 — Current standing-wave distribution on the antenna and matching section at 18 MHz. The antenna functions as two full-wave antennas, slightly folded up at the center, fed in phase.

section, the system loads well when used in conjunction with a suitable matching network and radiates effectively for DX contacts. See Fig. 6.

24 MHz: The antenna again functions effectively as a $5\lambda/2$ long wire, but because of the shift in the positions of the current antinodes on the flat-top and the matching section (Fig. 7), the matching or make-up section now presents a much lower resistive load condition to the feeder connected to its lower end than it does on 21 MHz. Again, the polar diagram is multilobed with low zenithal angle radiation.

28 MHz: On this band, the antenna functions as two long-wire antennas, each of three half waves, fed in phase. The polar diagram is similar to that of a $3\lambda/2$ long-wire, but with even more gain over a $\lambda/2$ dipole because of the collinear effect obtained by feeding two $3\lambda/2$ antennas, in line and in close proximity, in phase. See Fig. 8.

Construction

The Antenna

The dimensions of the antenna and its matching section are shown in Fig. 9. If possible, the flat-top should be horizontal and run in a straight line, and should be erected as high as can be above ground. In describing the theory of operation, it has been assumed that it is generally possible to erect the antenna at an average height of about 34 ft, which happens to be the optimum height for the antenna at 14 MHz. Although this is too low for optimum radiation efficiency on 1.8, 3.5, and 7 MHz for **any** horizontal type of antenna, in practice few amateurs can install masts of the optimum height of

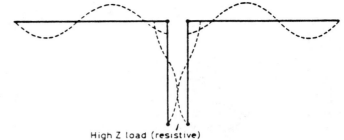

High Z load (resistive)

Fig. 6 —Current standing-wave distribution on the antenna and matching section at 21 MHz. On this band the antenna works as a long wire of five half waves. The base of the matching section presents a virtually nonreactive high impedance load to the feeder.

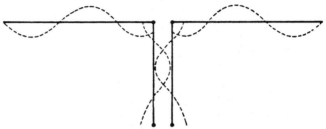

Resistive load approx 90/100 Ω

Fig. 7 — Current standing-wave distribution on the antenna and matching section at 24 MHz. The antenna functions as a long wire of five half waves.

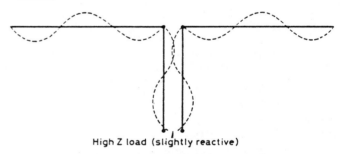

High Z load (slightly reactive)

Fig. 8 — Current standing-wave distribution on the antenna and matching section at 28 MHz. The antenna functions as two long-wire antennas each of three half waves length, fed in phase. A very effective form of antenna giving good multilobe, low zenithal angle, radiation.

half a wavelength at 3.5 or 7 MHz, and certainly not at 1.8 MHz.

If it is not possible to accommodate the 102-ft top in a straight line because of space limitations, up to about 10 ft of the antenna wire at each end may be allowed to hang vertically or at some convenient angle, or be bent in the horizontal plane, with little practical effect on performance. This is because, for any resonant dipole antenna, most of the effective radiation takes place from the center two-thirds of its length where the current antinodes are situated. Near each end of such an antenna, the amplitude of the current standing wave falls rapidly to zero at the outer extremities; consequently, the effective radiation from these parts of the antenna is minimal.

The antenna may also be used in the form of an inverted V. However, it should be remembered that for such a configuration to radiate at maximum efficiency, the included angle at the apex of the V should not be less than 120 degrees.\3 The use of 14 AWG enameled copper wire is recommended for the flat-top or V, although thinner gauges such as 16 or even 18 AWG can be used.

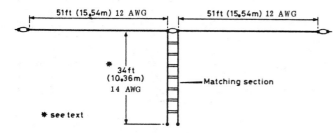

Fig. 9 — Construction dimensions of the G5RV antenna and matching section.

The Matching Section

This should be, preferably, of open-wire feeder construction for minimum loss. Since this section **always** carries a standing wave of current (and voltage), its actual impedance is unimportant. A typical, and satisfactory, form of construction is shown in Fig. 10. The feeder spreaders may be made of any high-grade plastic strips or tubing; the clear plastic tubing sold for beer or wine syphoning is ideal.

If you decide to use 300-ohm ribbon type feeder for this section, it is strongly recommended that the type with "windows" be used. It has lower loss than a feeder with solid insulation throughout its length, and it possesses relative freedom from the detuning effect caused by rain or snow. If this feeder is used for the matching section, allowance must be made for its velocity factor (VF) in calculating the mechanical length required to resonate as a half-wave section **electrically** at 14.15 MHz. Since the VF of standard 300-ohm ribbon feeder is 0.82, the **mechanical** length should be 28 ft. However, if 300-ohm ribbon with windows is used, its VF will be almost that of open-wire feeder, say 0.90, so its **mechanical** length should be 30.6 ft.

This section should hang vertically from the center of the antenna for at least 20 ft or more if possible. It can then be bent and tied off to a suitable post with a length of nylon or terylene cord at an above-head height. Supported by a second post, its lower end is connected to the feeder.

The Feeder

The antenna can be fed by any convenient type of feeder provided always that a suitable type of matching network is used. In the original article describing the G5RV antenna, published in the **RSGB Bulletin** for November 1966, it was suggested that if a coaxial cable feeder was used, a balun might be employed to provide the necessary unbalanced-to-balanced transformation at the base of the matching section. This was because the antenna and its matching section constitute a **balanced** system, whereas a coaxial cable is an **unbalanced** type of feeder. However, later experiments and a better understanding of the theory of operation of the balun indicated that such a device was unsuitable

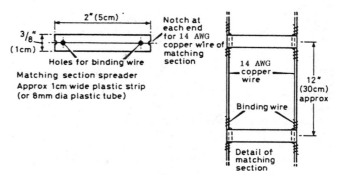

Fig. 10 — Constructional details of the matching section. Also suitable for open-wire feeder construction.

because of the highly reactive load it would see at the base of the matching or make-up section on most HF bands.

If a balun is connected to a reactive load with an SWR of more than 2:1, its internal losses increase. The result is heating of the windings and saturation of its core, if one is used. In extreme cases with relatively high power operation, the heat generated in the device can cause it to burn out. The main reason for not employing a balun in the G5RV antenna, however, is that unlike a matching network, which employs a **tuned circuit**, the balun cannot compensate for the reactive load condition presented to it by the antenna on most of the HF bands, whereas a suitable type of matching network can do this most effectively and efficiently.

Experiments were conducted to determine the importance, or otherwise, of unbalance effects caused by the direct connection of a coaxial feeder to the base of the matching section. There was a rather surprising result. The research showed that the HF currents measured at the junction of the inner conductor of the coaxial cable with one side of the (balanced) matching section, and at the junction of the outer coaxial conductor (the sheath) with the other side of this section, are virtually **identical** on all bands up to 28 MHz, where a slight, but inconsequential difference in these currents has been observed. There is, therfore, no need to provide an unbalanced-to-balanced device at this junction when using a coaxial feeder.

The use of an **unbalanced-to-unbalanced** type of matching network between the coaxial output of a modern transmitter (or transceiver) and the coaxial feeder is essential. This is because of the reactive condition presented at the station end of this feeder, which on all but the 14-MHz band, will have a fairly high to high SWR on it. The SWR, however, will result in insignificant losses on a good-quality coaxial feeder of reasonable length; say, up to about 70 ft. Either 50- or 80-ohm coaxial cable can be used. Because it will have standing waves on it, the actual characteristic impedance of the cable is unimportant.

Another convenient feeder type that can be employed is 75-ohm twin lead. It exhibits a relatively high loss at frequencies above 7 MHz, however, especially when a high SWR is present. I recommend that not more than 50 to 60 ft of this type be used between the base of the matching section and the matching network. The 75-ohm twin lead available in the United Kingdom is of the **receiver** type; less lossy **transmitter** type is available in the United States.

By far the most efficient feeder is the open wire type. A suitable length of such can be constructed in the same manner as that described for the open-wire matching section. If this form is employed, almost any length may be used from the center of the antenna to the matching network (balanced) output terminals. In this case, the matching section becomes an integral part of the feeder. A convenient length of open-wire feeder is 84 ft. It permits parallel tuning of the matching network circuit on all bands from 3.5 to 28 MHz, and with conveniently located coil taps in the matching network coils for each band, or where the alternative form of a matching network employing a three-gang 500 pF/section variable coupling capacitor is used, the optimum loading condition can be obtained for each band.\4 This is not a rigid feeder-length requirement, and almost any mechanically convenient length may be used. Since this type of feeder will always carry a standing wave, its characteristic impedance is unimportant. Sharp bends, if necessary, may be used without detriment to its efficiency. It is only when this type of feeder is correctly terminated by a resistive load equal to its characteristic impedance that such bends must be avoided.

Coaxial Cable HF Choke

Under certain conditions a current may flow on the **outside** of the coaxial **outer conductor.** This is because of inherent unbalanced-to-balanced effect caused by the direct connection of a coaxial feeder to the base of the (balanced) matching section, or to pickup of energy radiated by the antenna. It is an undesirable condition and may increase the chances of TVI [from fundamental overload, **if** the feeder is routed near a TV receiving antenna — Ed.]. This effect may be reduced or eliminated by winding the coaxial cable feeder into a coil of 8 to 10 turns about 6 inches in diameter immediately below the point of connection of the coaxial cable to the base of the matching section. The turns may be taped together or secured by nylon cord.

It is important that the junction of the coaxial cable to the matching section be made thoroughly waterproof by any of the accepted methods. Binding with several layers of plastic insulating tape or self-amalgamating tape and then applying two or three coats of polyurethane varnish, or totally enclosing the end of the coaxial cable and the connections to the base of the matching section in a sealant such as epoxy resin are a few methods used.

References

\1 Varney, L., "ATU or astu?," **Radio Communication,** August 1983.

\2 See Ref. 1.

\3 Varney, L., "HF Antennas in Theory and Practice — A Philosophical Approach," **Radio Communication,** Sept. 1981.

\4 See Ref. 1.

Vertical Antennas

Wiring Up the Old Spruce

By Kris Merschrod,* KA2OIG

*123 Warren Road, Ithaca, NY 14850

One serious problem an aspiring 80-meter operator may have is acquiring an antenna that will offer broadband performance and a low angle of radiation. I've been a Novice for two years, but early on discovered that the challenge of "getting out" on 80 meters had turned me into an antenna enthusiast. Friends suggested that only by going to a vertically polarized antenna could I expect to make it on 80 meters. That meant a vertical antenna. One characteristic of vertical antennas in articles I referenced is the need for an extensive radial system.[1] A second characteristic is the need for a matching system if the height is significantly less than 1/4 wave. If height was necessary to avoid using a matching system, the idea of a vertical was out of the question. The zoning regulations in my town were written to prohibit commercial radio antennas by establishing a 35-foot maximum antenna height. According to the articles in **QST**, this is a constitutional issue, and I'd rather spend my time on antennas than at the Supreme Court.

Although I'm currently limited to the Novice section of the band, broad bandwidth was not an immediate concern. Why put so much effort into an antenna of limited future? I read the articles on cage dipoles, but decided one of those "structures" would only invite comments from neighbors without solving the problem of the angle of radiation.[2, 3]

Finally, Wilson published an inspiring and straightforward article on a vertically polarized antenna with only two radials![4] I built it immediately and hung it from dipole supports up 35 feet. It provided excellent results; I could copy stations that couldn't be heard with my former dipole. It was not designed for local work, however, and I had complaints from the XYL (KA2OXG) because it was hanging in front of a main doorway. I carefully rolled up the "K4YF Special" for other locations; it's a great field day antenna.

I was hooked on vertical polarization, though, and resumed reading about this type of antenna with a little more skepticism concerning the number of radials really needed. One side remark in **The ARRL Antenna Anthology** on 80-meter phased arrays follows the usual admonition to keep the antenna free of obstacles and in the clear.[5] The comment was in relation to the swamp site of the array and an advertisement of the system by the COL-ATCH-CO.[6] Their ad stated that, "Foliage absorption at 40 and 80 meters is negligible." Their video tape available for loan was an opportunity to see the foliage in the swamp. Except to tie one end of the dipole, I had been avoiding the windbreak of tall spruces on the south side of my house for two years. The idea of a "hidden" antenna began to jell.

At the local scrap recycling yard, I obtained coils of cable TV feeder lines 3/4 inch in diameter. They are made of a fairly stiff aluminum-walled tube, filled with foam and a single copper conductor. At 50 cents a pound, I purchased them for a dime a foot. This chain of events convinced me to wire a 60-ft spruce in the following manner.

The Antenna Specifications

Spruce trees are like ladders, they have nicely spaced branches. A 60-ft length of TV feeder cable was laid on the ground (234/f (MHz) = 62.4 ft; I rounded the figure). This was pushed up the tree along the trunk until it began to bend over. Then I climbed the branches and pulled the cable along with me. Once at the 50-ft level, I slid the remainder of the "element" up ahead and lashed the cable to the tree. On the way down, I straightened the cable. After testing, the antenna was lashed at three levels to keep the lower end about 6 inches above the end of the central grounding rod.

The Radial System

An 8-ft ground rod was driven at the foot of the tree. Two 65-ft radials from the K4YF Special plus four 60-ft lengths of no. 8 aluminum were secured to the clamp. All of the radials are under the spread of the spruce trees so the "system" is about 65-ft east and west of the base, and 6 ft north and south.

The base feed point of the spruced-up vertical system for 80 meters.

Results

The noise bridge showed the impedance to be 80 ohms at the base of the antenna and a bit on the capacitive side (it could be shortened). A ratio of 1.2:1 from 3.5 MHz to 4.0 MHz was measured on the SWR meter. The feed line is a 52-ohm coaxial cable which is not a perfect match, but the SWR is pretty good. Thus, without a matching network, no capacitance hats, and a limited radial system, a 1/4-wave vertical put me on the air.

Using this antenna in the Novice band, I find that, like the K4YF Special, I can hear more than I can work. The stations that I can hear in the west or in the deep south probably can't hear me — at least they do not respond. Texas is as far as I have worked. One major problem is that most operators on the Novice segment of 80 meters work from dipoles, and distant signals are overcome by local stations. With my vertical, local stations are much weaker than they were on my dipole; thus, the distant stations are audible in comparison. My rig is a Drake TR-4C which puts in about 250 watts to the final, and I am consistently given 579s or better.

DXing has eluded me so far, but I think that the problem is with the Novice band allocation. Devoldere's booklet on 80-meter DXing shows that only Argentina, Chile, Canada, the Marshall Islands and the U.S. territories overlap CW sections with the U.S. Novice section (1978: p. 4-1).\7 These countries are a distance away, plus they may avoid the Novice segment because of the activity.

Outside of the Novice band, reception is great. It is fun to listen in the Extra-Class segment in the so-called DX window. I've heard some European and Japanese stations in pileups never experienced on the Novice segment.

Conclusion

What would hams do without trees? Sounds like a good opener for a joke, but on 80 meters and below, trees for antenna supports may be the easiest method to get on the air. The neighborhood looks the same and the zoning board isn't bothered. As luck would have it, there are other spruces a quarter wave away east and west. I'm planning on sprucing up the next phase of my antenna experiments.

The vertical radiator (left) snakes its way up the trunk of the spruce tree, where it is lashed at the top.

References

\1 Sevick, Jerry, W2FMI, "The Ground-Image Vertical Antenna," **The ARRL Antenna Anthology,** 1978, (ARRL: Newington, CT), p. 22.

\2 Hall, Jerry, K1TD, "The Search for a Single Broadband 80-Meter Dipole," **QST,** April 1983, pp. 22-27.

\3 Wolfgang, Larry, WA3VIL, "West Jersey Communication Products 80-Meter 'BN Cage' Antenna," **QST,** Sept. 1983, pp. 43-44.

\4 Wilson, John J., K4YF, "The 'K4YF Special' Antenna," **QST,** Sept. 1978, pp. 26-27.

\5 Atchley, Dana W., "Switchable 4-Element 80-Meter Phased Array," **QST,** March 1965; **The ARRL Antenna Anthology,** 1978, pp. 119-122.

\6 COL-ATCH-CO, advertising folder for "Instantarry Systems," P. O. Box 230, Carlisle, MA 01741, 1983.

\7 Devoldere, John, ON4UN, **80-Meter DXing,** Communications Technology, Inc., Greenville, NH 03048, 1978.

A Triband Parasitic Vertical Directional Array

By Walter J. Schulz,* K3OQF

*3617 Nanton Terrace, Philadelphia, PA 19154

Vertical directional arrays are employed in AM broadcast stations to protect other radio stations that share the same frequency from interference. In other instances, they are used to direct radiation to a specific target area. It is with that purpose in mind that as radio amateurs we are interested in this type of array. Much of this work was completed during the 1930s on the broadcast band in the United States, based on the research of Dr. George H. Brown and Carl E. Smith.[1,2]

We have occasionally seen directional arrays in Amateur Radio articles with all the vertical elements driven. Sometimes the array is constructed with one driven element and one parasitic element. This type of array is a much simpler system to employ. However, when the number of array vertical elements is increased beyond two, the difficulty increases. I was perplexed on how to solve this problem in a simple way and felt that by using an array where each element was driven would prove a difficult task to maintain in adjustment, especially with an inadequate ground radial system. One of the other associated problems is sustaining the correct phase relationships and current ratios between array elements. Therefore, something less complicated was needed to overcome the problems noted. The answer occurred to me in the form of a triband Yagi-Uda dipole array conforming to the configuration of a vertical triband parasitic array placed on the earth's surface.

Directional Array with Parasitic Elements

A quarter-wave element can be constructed by dividing a Yagi-Uda array element in half. By employing the earth's surface with a good ground radial system, an image quarter-wave antenna is produced under the earth's surface. Using our imagination, we can visualize that the image element exists along with the real element above the earth's surface. The image and real element produce a half-wave element whose feed point is at the earth's surface. Such a Yagi-Uda imaginary configuration would be vertically polarized. Constructing such a directional array in real life would be a good compromise, offering good forward gain, possessing a fairly good front-to-back ratio to discriminate against unwanted signals, and having the capability to operate over the 20-, 15-, and 10-m radio amateur bands.

The goal then is to design a triband parasitic vertical array consisting of one driven element, one reflector, and one director. The antenna system must endfire to the target area and be fed with a coaxial transmission line having a 52-ohm characteristic impedance. Judicious selection of spacing is required to maintain an inductive or capacitive reactance at the feed point over the three bands of operation. I decided to keep the reactance capac-

itive and employ a T-matching network using variable inductors in its arms. In this manner, you can tune for an impedance match in your station, provided the resonant traps on the driven element (dielectric) do not break down from the voltage standing waves. To find the right values for these electrical parameters, a certain amount of perturbation was necessary. The electrical parameters were obtained with a pocket calculator such as the Hewlett-Packard HP-15C. By proper selection of self- and mutual-impedance values from known charted values, current magnitudes and phases on each element were found. Then, from these electrical current values, the drive-point impedance and radiation pattern over the three bands were found.

Algorithm for Finding Currents in Array Elements

Before we discuss a keystroke sequence that will yield the current value for each individual element in the array, let us look at some diagrams. We know from observing a directional array that by making the observation directly above the system, an analogy to isotropic point sources can be seen.[3] It becomes useful to represent the triband parasitic directional array in this manner for each band of operation. Notice for each illustration shown in Fig. 1A, B and C, that the electrical spacing is represented in electrical degrees. Between elements, the spacing becomes larger electrically as the wavelength becomes shorter, although the physical distance between the elements remains the same. This occurs because the wavelength changes for each band of operation.

We begin by illustrating the problem graphically, finding the mutual impedance representative of the physical spacings for each band of operation. The driven element height is approximately 0.236 wavelength, the reflector element is 0.247 wavelength, and the director element is 0.225 wavelength. The antenna uses a 1-inch OD aluminum tubing. The self-impedance value is based on these forementioned heights and the outside diameter throughout the design. Our next step, once the self- and mutual impedances are obtained, is to calculate the currents flowing in the array.

It is helpful to think of each vertical element in terms of self-impedance, and one that forms a network. When we think of a network, the mutual impedances must be considered also. Our goal is to solve for the current magnitude and phase of each mesh that represents a vertical element in the array or network.[4] We may start by writing the simultaneous equation describing the problem:

$$1 = I_1 \times Z_{11} + I_2 \times Z_{12} + I_3 \times Z_{13}$$

$$0 = I_1 \times Z_{21} + I_2 \times Z_{22} + I_3 \times Z_{23}$$

$$0 = I_1 \times Z_{31} + I_2 \times Z_{32} + I_3 \times Z_{33}$$

Notice that the parasitic elements have no voltage flowing in the mesh, but the driving voltage is generated in the first equation. This driving voltage will induce current flow in the other meshes. We must determine the current magnitude and phase for I1, I2, and I3 before we can determine the drive-point impedance and the radiation patterns. There are a number of systems that will yield these current values. Determinants or Guassian Successive Elimination methods are commonly used. However, with the advent of the pocket calculator, and particular the HP-15C, these methods are already programmed in the calculator. This reduces the design work to just plugging in the values. The two mathematical systems used to solve for currents in the array are included in the appendix.

The Matrix

Let's consider the problem described above for 20 meters and use the electrical parameters below for this given band of operation.

Self-Impedances:

Z_{11} 31 −j8   driven element

Z_{22} 36 −j21 reflector element

Z_{33} 27 −j37 director element

Mutual Impedances:

$Z_{12}, Z_{21} = 36 + j7$

$Z_{13}, Z_{31} = 36 + j7$

$Z_{23}, Z_{32} = 29 - j6$

We can then write the complex impedances in terms of individual entries in a matrix along with the voltages.

$$\begin{bmatrix} Z_{11} & Z_{12} & Z_{13} \\ Z_{21} & Z_{22} & Z_{23} \\ Z_{31} & Z_{32} & Z_{33} \end{bmatrix} \times \begin{bmatrix} I_1 \\ I_2 \\ I_3 \end{bmatrix} = \begin{bmatrix} 1 \\ 0 \\ 0 \end{bmatrix}$$

Matrix A   Unknown   Matrix B

Next, enter the complex values into the individual entries.

Matrix A   Matrix B

$$\begin{bmatrix} (31 - j8) & (36 + j7) & (36 + j7) \\ (36 + j7) & (36 + j21) & (29 - j6) \\ (36 + j7) & (29 - j26) & (27 - j37) \end{bmatrix} \times \begin{bmatrix} I_1 \\ I_2 \\ I_3 \end{bmatrix} = \begin{bmatrix} 1 + j0 \\ 0 + j0 \\ 0 + j0 \end{bmatrix}$$

Because the impedance values are complex (rectangular form), we must determine a method to write the impedance matrix in such a way that the HP-15C calculator can handle the matrix. This is accomplished by partitioning the matrix into two submatrices, one having a real-number entry, while the other submatrix has the imaginary (j-operator) value entered. We write the new matrix as:

Matrix A   Matrix B

$$\begin{bmatrix} 36 & 36 & 36 \\ 36 & 36 & 29 \\ 36 & 29 & 27 \\ \hline -8 & 7 & 7 \\ 7 & 21 & -6 \\ 7 & -6 & -37 \end{bmatrix} \times \begin{bmatrix} 1 \\ 0 \\ 0 \\ \hline 0 \\ 0 \\ 0 \end{bmatrix}$$

Notice that the submatrix entry positions are identical; the associated imaginary number entries are identical with the entry positions in the upper submatrix. It is important to keep the entries in the correct positions in both submatrices. Next to be considered is Matrix B. This matrix must be written for the voltage values in the same manner. Remember that this matrix is also partitioned and it may prove helpful for you to write one column at a time with its real and associated imaginary numbers.

We are now ready to proceed with the actual keystroke procedure as described in the HP-15C owner's manual.\5.6

Keystroke Sequence

```
0 f DIM (i)
f MATRIX 1
6 Enter 3 f DIM A (size of Matrix A is 6 x 3)
f MATRIX 1
f USER
31 STO A 1,1
36 STO A 1,2
36 STO A 1,3
36 STO A 2,1
36 STO A 2,2
29 STO A 2,3
36 STO A 3,1
29 STO A 3,2
27 STO A 3,3
-8 STO A 4,1
7 STO A 4,2
7 STO A 4,3
7 STO A 5,1
21 STO A 5,2
-6 STO A 5,3
7 STO A 6,1
-6 STO A 6,2
-37 STO A 6,3
6 ENTER 1 f DIM B
1 ENTER 1 ENTER
STO g (B)
RCL MATRIX B 6 1
RCL MATRIX A 6 3
f MATRIX 2
A 6   6
f RESULT C
%
C6      1
g Cy,x (converts partition to complex form −
                            rectangular)
I_1 = RCL C Re 1,1 0.038
      RCL C j 1,2 0.015
I_2 = RCL C Re 2,1 −0.0.36
      RCL C j 2,2 0.027
I_3 = RCL C Re 3,1 0.027
      RCL C j 3,2 −0.030
```

The Drive-point Impedance

Using the above method, we are able to derive the current magnitude and phase for each element in

95

the antenna array. Next, we must determine two electrical values: the drive-point impedance and the radiation pattern. The equation shown below gives the drive-point impedance for the array, provided we know the current ratios between the driven element to the reflector and the director elements.

$$z_{in} = z_{11} + I_2/I_1 \,(z_{12}) + I_3/I_1 \,(z_{13})$$

$$z_{in} = (31 - j8) + (-0.036 + j0.027)/0.038 + j0.015 \,(36 + j7) \\ + (0.027 - j0.030)/0.038 + j0.015 \,(36 + 7)$$

$$z_{in} = (31 - j8) + (-27.338 + j29.729) + (18.902 - j30.921)$$

$$z_{in} = 22.564 - j9.192 \text{ ohms}$$

The procedure for finding the element currents and drive-point impedances must be completed for each band of operation. This must be done because the mutual impedances will change between elements with different electrical spacings.

The Radiation Pattern

Refer to Figs. 1A, B and C. The radiation pattern is determined from the current ratios between the elements. Take a closer look at the drive-point impedance equation and observe the two ratios of interest: I2/I1 and I3/I1. When the drive-point impedance is divided by itself, it will be equal to 1 + j0; when converted to polar form it is 1.0/ 0 . Now we must divide the current in complex rectangular form, I1, into I2 and I3, and then convert the quotient to polar form for manipulation. It becomes evident that there are now three polar form values that describe the magnitude and phase between elements. Further, by considering the physical spacing between elements, the radiation pattern field strength intensity is determined for each azimuth angle. The easiest method to write the vectors is in the following form:

Magnitude /spacing cos θ − θ1 + electrical phase

where

    θ = azimuth angle
    θ₁ = azimuth position of vertical element

The vectors are written using the chain method, which is simply the addition of all vectors graphically.\7,8 Once all the vectors have been added, the resultant magnitude is obtained for a specific azimuth angle that is under consideration. The chain, also known as the head to tail method, is repeated again until all azimuth angles in the pattern have been computed.

Once the horizontal radiation pattern has been obtained (Figs. 2A, B and C), the gain figure over the isotropic point source is derived by finding the integral of the calculated radiation pattern. There are a number of methods that can be employed, but the simplest is that of the solid angle concept, borrowed from optics. To use this concept, we must know the beamwidth in both the horizontal and vertical planes. The 0.707 root-mean-square value is multiplied by the maximum field strength intensity in both planes. The product is the half-power point in the radiation pattern. This value is found along the radiation pattern plot. Once it is located, note the azimuth angle. There will always be two half power points in the horizontal plane, while there will be only one at a specific elevation angle in the vertical plane. Once we know the beamwidth, the equation below is used to find directivity and power gain in decibels.

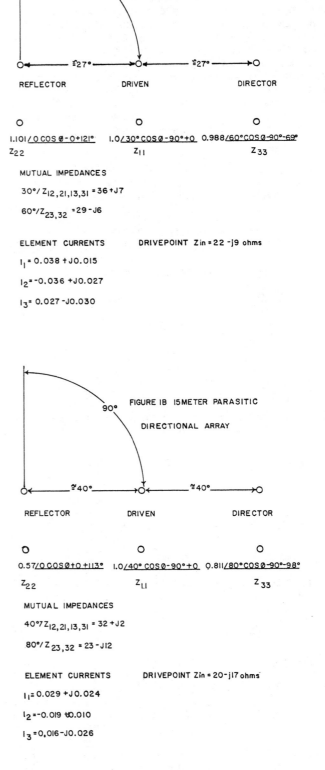

Fig. 1 -- Self- and mutual-impedance values taken from known charted values were employed to calculate current magnitudes and phases on each element. From these electrical current values, the drive-point impedance and radiation pattern for each was found.

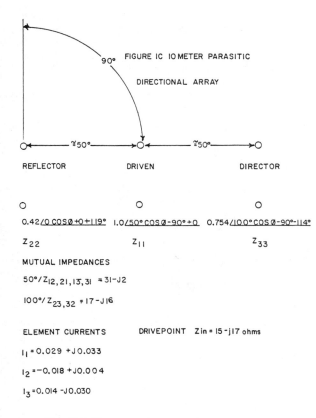

90°

FIGURE IC 10METER PARASITIC

DIRECTIONAL ARRAY

≈50°   ≈50°

REFLECTOR   DRIVEN   DIRECTOR

$0.42 / \underline{0 \cos\emptyset + 0 + 119°}$ $1.0 / \underline{50° \cos\emptyset - 90° + 0}$ $0.754 / \underline{100° \cos\emptyset - 90° - 114°}$

Z_{22} Z_{11} Z_{33}

MUTUAL IMPEDANCES

$50° / Z_{12,21,13,31} \cong 31 - j2$

$100° / Z_{23,32} \cong 17 - j16$

ELEMENT CURRENTS   DRIVEPOINT $Z_{in} = 15 - j17$ ohms

$I_1 = 0.029 + j0.033$

$I_2 = -0.018 + j0.004$

$I_3 = 0.014 - j0.030$

$$D = 41,253 / (\Theta_v \times \Theta_h) \qquad \text{and}$$

$$\text{dB gain} = 10 \log_{10} D$$

In our particular case the system under design will yield a higher value than can actually be expected in the real world. However, the technique serves the purpose of allowing us to obtain a better understanding of system gain for the array. It helps to evaluate our design goals — and to see whether we are on the right path.

Traps and Loading

We have designed a 3-element triband parasitic directional array, but now we must calculate the effect that resonant LC trap loading will have on each element length. Not only does a resonant trap on its operation frequency isolate the antenna element beyond it, but when the trap is operated lower in frequency, it will effectively load the antenna element inductively. This loading effect makes the antenna element longer electrically, but also makes it possible for shorter element length (the actual physical dimension). It is evident from the preceding discussion that we must consider trap inductive loading in our design.

In recent years, Joseph Boyer, W6UYH, wrote a series of articles describing how to calculate and design trap verticals. This knowledge is applied here to design the reflector, driven and director elements.\9,10,11,12 However, there is one slight difference. Rather than assume each element is of quarter-wave height, note in Fig. 3A that each element height is different. This is important because each length-to-diameter ratio is different; that means the shape factor is different. All values are used with this shape factor in our calculations on each element. From the shape factor, it is possible to consider how element and trap series inductive reactance adds to the element length

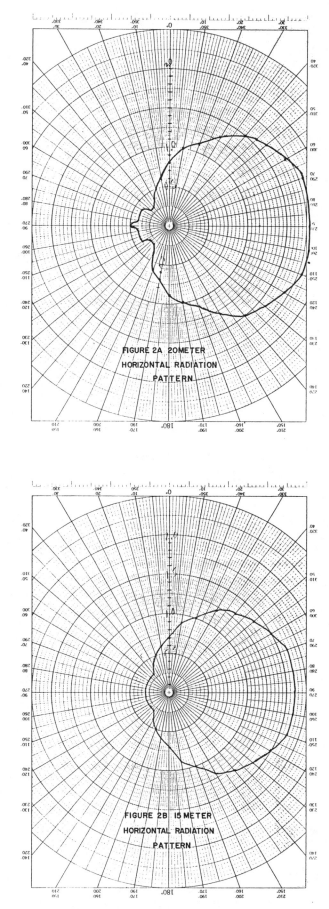

FIGURE 2A 20METER
HORIZONTAL RADIATION
PATTERN

FIGURE 2B 15METER
HORIZONTAL RADIATION
PATTERN

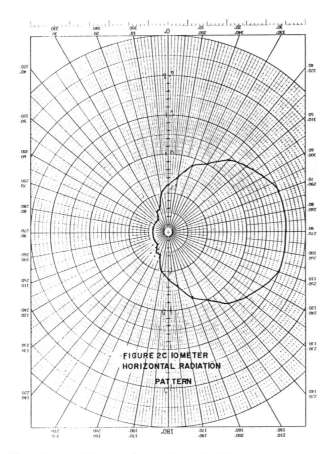

Fig. 2 — The horizontal radiation patterns for each band are shown. At A, we have the 20-m pattern, B depicts the 15-m band, and C is the 10-m band.

electrically. The net result of this is that the element physical height is shortened.

Our procedure is to arbitrarily select a fixed capacitance for the trap and then use its reactance value to calculate the inductance value to make the trap resonant. Refer to Fig. 3B. Following step by step, we must determine the loss resistance of the inductance, $(R_L = X_L/Q)$, and the impedance of the LC constants in the trap at its resonant frequency:

$$(Z_p = X_L^2/R_L)$$

Finally, we must find the value of series reactance that the trap presents as an inductive loading when operated on frequencies other than resonance.

$$X_S = (1/Q \, M \, (1/M^2 - 1)) \, (Z_p)$$

$$M_{(10)} = 21.225/28.6 = 0.742$$

$$M_{(15)} = 14.175/28.6 = 0.496$$

$$M_{(20)} = 14.175/21.225 + 0.668$$

Shape Factor

Let's briefly return to the subject of shape factor. Dr. S. A. Schelkunoff gave a reliable equation for shape factor, which is sometimes referred to as Ka or Z⊖\13 He equated an antenna to a transmission line in its behavior and this enabled him to find the antenna feed point complex impedance. What interests us here is that on each

band of operation we must describe the length-to-diameter ratio of each element height. The length-to-diameter ratio has already been used in the calculation of element self impedance, but for those who are not familiar with the procedure, I use the equation by Schelkunoff:

$$Z_\ominus = 60 \, (\log_e 2h/a - 1)$$

where
    a = radius
    h = height

Not only does this shape-factor equation determine the electrical height, but will enable us to determine the voltage potential at each trap.

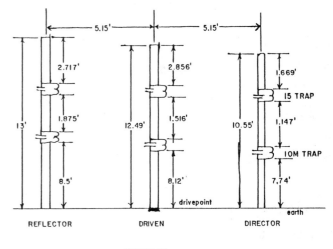

FIGURE 3A

TRIBAND PARASITIC DIRECTIONAL ARRAY

Fig. 3A — Note that each element height varies. This is important because each length-to-diameter ratio is different, thus the shape factor also varies.

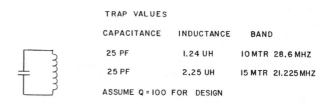

| TRAP VALUES | | |
CAPACITANCE	INDUCTANCE	BAND
25 PF	1.24 UH	10MTR 28.6 MHZ
25 PF	2.25 UH	15 MTR 21.225MHZ
ASSUME Q = 100 FOR DESIGN		

FIGURE 3B TRAP LC VALUES

Fig. 3B — Trap LC values for the 10- and 15-meter band.

Voltage Potential on the Trap

Practical experience has shown that a large amount of voltage is present on the upper element portion of a vertical antenna. For short vertical heights less than one quarter wave, the maximum current flows at the antenna base while high voltage potential is a cosine function in regard to antenna height. What concerns us the most in the configuration described is the voltage potential across the traps. The trap components must have significant dielectric strength to prevent breakdown from either tuning standing wave voltages or just daily usage.

We begin by finding the quality factor of the driven element.[14]

$$Q_{(20)} = Z_\theta - 3/46.5 = 295.593 - 3/46.5 = 6.3$$

From calculations performed, we know that the array exhibits a radiation resistance of 22 ohms on the 20-m band. We assume the reactive component will be canceled by the series capacitance arm of the T network. Finding the current at the array drive point for an output power of one kilowatt:

$$I = (p/Rr)^{0.5} = (1000/22)^{0.5} = 6.7 \text{ amperes}$$

The drive-point voltage is determined next:

Voltage at drive point = I Rr = (6.7)(22) = 148.3 V and from the product of voltage times the quality factor (Q) we find:

Voltage at trap = Q V/2 = (6.3)(148.3)/2 = 467.2 V

It is advisable to use a safety factor of approximately four times the calculated voltage value. This is satisfactory with materials available today. Also, I suggest the use of half-wavelength multiples of coaxial transmission line to feed the array.

The traps may be constructed from other than lumped-constant component parts and it is probably cheaper to employ coaxial LC traps. Not only is it less expensive, but the dielectric strength will be greater depending on the coaxial cable chosen.

Some last words about trap construction. It may be easier to construct coaxial traps first, and then measure their quality factor with a Q meter. Also, the inductive reactance and coil loss resistance can be calculated. Once this is accomplished, parallel resonant impedance may be determined. Finally, from those values, the series reactive loading of the trap is found. By changing the procedure, we are still able to calculate the correct linear antenna height in each element.

This article is a brief outline of a design for a triband parasitic directional array. It has significant advantages in that by doubling the element identical we have a blueprint for a Yadi-Uda dipole array. In this case, the drive-point complex impedance will double in value and a matching network will have to be employed for matching the transmission line to the antenna. Using the array as a directional array, employ a ground radial system to prevent loss resistance and maintain a close resemblance to the theoretical radiation pattern shown in Figs. 2A, B and C for each band.

References

[1] Brown, G. H., "Directional Antennas," **Proceedings of the Institute of Radio Engineers,** Jan. 1937, p. 78-145, vol. 25, no. 1, part 1.

[2] Smith, Carl, **Directional Antenna Patterns,** Cleveland: Cleveland Institute of Radio Electronics, 1958.

[3] Schulz, W., "Vertical Array Analysis," **QST,** Feb. 1981, pp. 22-25.

[4] Schulz, W., "Key to Three Element Yagi Design," **Ham Radio,** March 1984, pp. 48-51.

[5] **Hewlett-Packard HP-15C Owner's Handbook,** Aug. 1982, pp. 160-173.

[6] **Hewlett-Packard HP-15C Advanced Functions Handbook,** Aug. 1982, pp. 128-131.

[7] Schulz, W., "HP-41C Tweaks Vertical Antenna Arrays," **Electronics,** Jan. 17, 1980, pp. 141-143.

[8] See Reference 3.

[9] Boyer, Joseph, "The Multiband Trap Antenna, Part 1," **CQ Magazine,** Feb. 1977, pp. 26-30 and 73-74.

[10] Boyer, Joseph, "The Multiband Trap Antenna, Part 2," **CQ Magazine,** March 1977, pp. 51-55 and 72.

[11] Boyer, Joseph, "The Multiband Trap Antenna, Part 3," **CQ Magazine,** April 1977, pp. 46-50 and 72.

[12] Boyer, Joseph, "The Multiband Trap Antenna, Part 4," **CQ Magazine, May 1977,** pp. 22-27.

[13] Schelkunoff, S.A., "Theory of Antennas of Arbitrary Size and Shape," **Proceedings of the Institute of Radio Engineers,** Sept. 1941, pp. 493-520.

[14] Schelkunoff, S.A., and Friis, Harold, **Antennas Theory and Practice,** New York: John Wiley & Sons, Inc., 1952, pp. 435-436.

[15] Kurosh, A., **Higher Algebra,** Moscow: Mir Publishers, 1980, pp. 15-22.

[16] See Reference 4.

Appendix

The Algorithm for Successive Elimination of Three Simultaneous Equations

$$I_1 Z_{11} + I_2 Z_{12} + I_3 Z_{13} = V1$$
$$I_1 Z_{21} + I_2 Z_{22} + I_3 Z_{23} = V2$$
$$I_1 Z_{31} + I_2 Z_{32} + I_3 Z_{33} = V3$$

The first step is to eliminate $I_1 Z_{21}$ and $I_1 Z_{31}$.

$$I_1 Z_{11} + I_2 Z_{12} + I_3 Z_{13} = V1$$
$$+ I_2 Z_{22}' + I_3 Z_{23}' = V2'$$
$$+ I_2 Z_{32}' + I_3 Z_{33}' = V3'$$

$$Z_{22}' = (Z_{22} - Z_{12} Z_{21}/Z_{11})$$
$$Z_{23}' = (Z_{23} - V_1 Z_{21}/Z_{11})$$
$$Z_{33}' = (Z_{33} - Z_{13} Z_{31}/Z_{11})$$
$$V_2' = (V_2 - V_1 Z_{21}/Z_{11})$$
$$V_3' = (V_3 - V_1 Z_{31}/Z_{11})$$
$$Z_{32}' = (Z_{32} - Z_{12} Z_{31}/Z_{11})$$

Gaussian pivot is Z_{21}/Z_{11} for the first-step elimination method.

Second Step

In the second step, the first simultaneous equation is no longer involved. We only manipulate the second and third equation using the following pivot Z_{32}'/Z_{22}'.

$$Z_{33}" = (Z_{33}' - Z_{23}' Z_{32}'/Z_{22}')$$

$$V3" = (V3' - V2'Z_{32}'/Z_{22}')$$

then:

$$I_1Z_{11} + I_2Z_{12} + I_3Z_{13} = V1$$
$$+ I_2Z_{22}' + I_3Z_{23}' = V2'$$
$$+ I_3Z_{33}" = V3"$$

finding in the last equation that I3 = V3"/Z33" and using this value for I3, we then back fill the second equation to find I2 = V2' - I3Z23'/Z22. Again using the values for I3 and I2, we back fill the first equation to find I1. It is in this manner that the solution to all unknown current values are found using the Gaussian successive elimination method.\16

The Second Method Called Determinants

A far more simpler method to find unknown current values is the use of determinants. This method is useful so long as the simultaneous equations are non-singular, or you can arrive at a determinant for the system.

$$I_1 = (Z_{22} Z_{33}) - (Z_{32})^2/DET$$

$$I_2 = (Z_{23} Z_{31}) - (Z_{33} Z_{21})/DET$$

$$I_3 = (Z_{21} Z_{32}) - (Z_{31} Z_{22})/DET$$

$$DET = (Z_{11} Z_{22} Z_{33} + Z_{12} Z_{23} Z_{31} + Z_{13} Z_{21} Z_{32} - Z_{31} Z_{22} Z_{13} - Z_{32} Z_{23} Z_{11} - Z_{33} Z_{21} Z_{12})$$

The 5/8-Wavelength Antenna Mystique

By Donald K. Reynolds,* K7DBA

*230 B St., No. 301, Friday Harbor, WA 98250

The 5/8-wave monopole antenna is well known to amateurs operating in the VHF/UHF mobile communications area. Often the operator is not an expert in electromagnetic theory or wave propagation. Because of this, many amateurs have the misconception that there must be something mysterious and superior about this particular length.

What is so magical about the 5/8-wave monopole? Actually, it is an old idea. The earliest publication on the subject (to my knowledge) appeared in the **Proceedings of the Institute of Radio Engineers** in 1924.\1

The purpose of this article is to review the operation of the 5/8-wave monopole. I will point out its desirable properties, as well as certain applications in which it is not only undesirable, but can considerably degrade the performance of a communications system.

Radiation Pattern and Gain

AM broadcasting was in its infancy in the early 1920s. Broadcast engineers were interested in determining the height of a base-excited vertical-tower radiator that would produce the maximum ground wave field strength for a given constant input power to the antenna. A technical paper published in 1924 showed, theoretically, that subject to certain assumptions, the tower height for maximum ground-wave field strength was 5/8 wavelength.\2 These are the assumptions:

1. Consider the tower to be a thin filament of perfectly conducting wire, mounted perpendicular to a plane of infinite extent and conductivity.

2. Consider the current amplitude to be a sinusoidal function of the distance down from the upper extremity of the antenna, as given by:

$$I(z) = I_m \sin\left[\frac{2\pi(\ell-z)}{\lambda}\right] \qquad (Eq.\ 1)$$

where I_m is the maximum current amplitude.

    z is the distance from the base to the point where I is calculated.

    ℓ is the total length of the antenna.

    λ is the wavelength, in the same units as z and ℓ.

The standing-wave current distribution we have assumed is shown in Fig. 1 for a 5/8-wave antenna. The relative phase of the assumed current, with respect to the phase of a voltage generator at the base, is also plotted. Note that the current in the upper 1/2 wavelength is of constant phase — 180 degrees out of phase with the current in the lower 1/8 wavelength. Also note that the amplitude of the current at the base (feed point) is 0.707 I_m. Since the input current leads the applied voltage by 90

degrees, the input impedance is a capacitive reactance.

Actual current distributions on wires and rods of finite diameter differ from this assumed distribution in certain details. The current on the wire may be broken down into two components; one in phase with the driving voltage, the other 90 degrees out of phase. The resultant total current is the phasor sum of the two components, and is plotted in dashed lines for a typical case in Figs. 1B and 1C. The phase of the total current no longer shifts abruptly from −90 to +90 degrees at the point 1/2 wavelength down from the upper end, but varies more gradually. Also, the magnitude of the current does not go to zero at the point 1/2 wavelength down from the top, but has a minimum value a little less than 1/2 wavelength down. The smaller the diameter of the antenna, the lower is this current minimum, but the diameter must be extremely small compared to the wavelength for the current minimum to approach zero. For example, if the antenna consists of a uniform cylindrical wire whose radius is 0.0001 wavelength, the current minimum is about 20% of the maximum value. For a radius of 0.01 wavelength, the minimum current is nearly 50% of the maximum value.

The amplitude and phase of the current along cylindrical monopoles of different lengths and diameters have been calculated using digital computers with what are called "moment methods."\3 Accurately measured current distributions reported by T. Morita show close correspondence with the more recently calculated distributions.\4

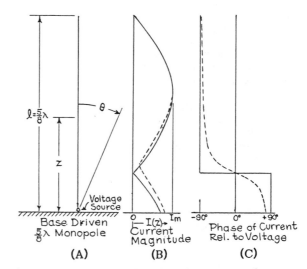

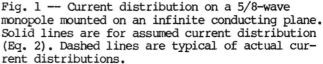

Fig. 1 — Current distribution on a 5/8-wave monopole mounted on an infinite conducting plane. Solid lines are for assumed current distribution (Eq. 2). Dashed lines are typical of actual current distributions.

Even though the actual current distribution on real 5/8-wave monopoles differ considerably from the ideal standing wave distribution given by Eq. 1, the radiation pattern computed on the basis of Eq. 1 is quite close to measured patterns. The expression for the radiation pattern, calculated from the current distribution of Eq. 1, is to be found in many textbooks, and is given by:[5]

$$E = \frac{0.5858 \left[\cos(225° \times \cos \theta) + 0.7071\right]}{\sin \theta} \quad \text{(Eq. 2)}$$

where E is the relative electric field strength at a large radial distance from the antenna.

θ is the vertical angle measured with respect to the axis of the antenna, as shown in Fig. 1A.

The maximum value of E occurs at θ = 90 degrees, and the factor of 0.5858 adjusts this value to unity, making Eq. 2 the "normalized E-plane pattern."

Fig. 2 shows the pattern of Eq. 2, expressed in decibels. I have also plotted the patterns of a 1/4- and a 1/2-wave monopole. All three patterns correspond to the same radiated power, with the maximum value of the 1/4-wave monopole taken as 0 dB. Note that the maximum radiated power occurs at θ = 90 degrees for all three monopoles. The 5/8-wave monopole is seen to give almost exactly 3 dB of gain over the 1/4-wave monopole, a fact which was observed by Ballentine.[6]

Fig. 3 shows the calculated theoretical power gains of vertical monopoles of different lengths mounted above an infinite conducting plane, with respect to the gain of a 1/4-wave monopole. All gains are calculated for θ = 90 degrees, that is, toward the horizon. This curve is derived from Terman.[7] Note that the gain of a very short monopole compared to the wavelength is only about 0.4 dB below that of 1/4-wave monopole.

For lengths less than 1/2 wavelength, the radiation pattern has a single lobe centered on the horizon. Above 1/2 wave, a minor lobe appears at a high angle, an effect which tends to reduce the gain toward the horizon. However, the main lobe at θ = 90 degrees continues to narrow, tending to increase the gain, with the result that maximum gain at θ = 90 degrees occurs for the well known length of 5/8 wavelength, with a value of almost exactly 3.0 dB with respect to the 1/4 wavelength. For lengths greater than 5/8 wavelength, the minor lobe rapidly increases in amplitude, and the gain in the direction θ = 90 degrees falls off, reaching zero (− ∞ dB) for a length of 1 wavelength.

The foregoing predicted radiation patterns and gains occur **only** when the following two requirements are satisfied:

1) The base of the antenna is located on a flat ground plane of infinite extent and infinite conductivity.

2) The observer is on the ground plane when measuring the maximum radiated field. The ground plane extends **beyond** the observer (theoretically to infinity).

These requirements are approximated reasonably well for vertical AM broadcast antenna towers (albeit with some degradation because of finite ground conductivity), since the receiving stations

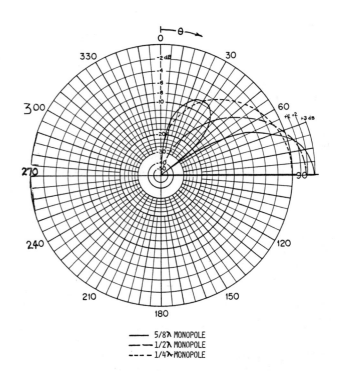

--- 5/8λ MONOPOLE
– – 1/2λ MONOPOLE
- - - - 1/4λ MONOPOLE

Fig. 2 — Theoretically calculated radiation patterns in vertical plane for monopoles on an infinite conducting plane. Zero dB corresponds to maximum radiation intensity of 1/4-wave monopole.

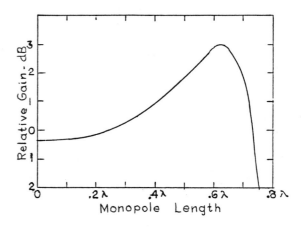

Fig. 3 — Power gain toward the horizon of a vertical monopole on an infinite conducting plane plotted vs length of monopole.

are usually at ground level, and the earth continues beyond the receiver. These requirements are **not** satisfied in the five common installations of 5/8-wave monopoles described below, with the result that the signal level radiated toward the horizon can be much less than the theoretical value.

Improper Uses of 5/8-Wave Monopoles

The following section describes the unhappy consequences of five common installations of 5/8-wave monopoles.

1) An antenna mounted at the end of a length of coaxial cable, with no ground plane of any kind.

In a rush to get the 2-meter FM base station on the air, operators sometimes screw on a commercial 5/8-wave whip, with a coaxial fitting on the base, to the end of a length of coaxial cable. The antenna is erected, possibly attached to a broom handle.

What we have now is actually a **dipole**; the upper element is the 5/8-wave whip, and the lower element is the **outer conductor** of the coaxial cable, extending back to the transceiver. The RF current on the center conductor of the coaxial line excites the whip. The current on the outer conductor, equal in magnitude to the current on the center conductor, simply folds over and excites the outside of the outer conductor below the feed point (the other element of the dipole).

The radiation pattern for the system varies with each installation, depending on such factors as the length of the coaxial cable, how it is dressed, proximity to other objects, and so on. The radiation intensity in the desired direction (toward the horizon) will almost certainly be severely degraded, compared to that of a properly designed vertical base-station antenna. Furthermore, there will probably be RF currents on the outside of the transceiver, power supply leads, and possibly the house wiring.

To illustrate these points, vertical-plane radiation patterns were made on a 5/8-wave commercial whip antenna for two feed distances: 1 wavelength and 1-1/2 wavelengths. The transceiver was connected to a length of coaxial cable that extended below the feed point in a straight line with the whip. These measurements were taken using the technique described in the last section of this article. The resulting patterns are plotted in dB vs vertical angle in Fig. 4, along with the pattern of a theoretical 5/8-wave monopole. Note the almost total lack of correspondence between the measured and theoretical patterns. Gains cannot be compared on these superimposed patterns, since all of them are normalized to 0 dB in the direction of maximum radiation for each antenna. However, it can be positively stated that since gain in a desired direction can only be obtained by reducing the radiation intensity in undesired directions, and redirecting it in the desired direction, the gains toward the horizon in the two measured cases must be much lower than in the theoretical case. Note particularly that in the case of the whip with one wavelength of coaxial line below, maximum radiation intensity occurs at an angle of about 25 degrees above the horizon, and drops 6 dB below that level in the desired direction of θ = 90 degrees.

2) An antenna mounted on a finite ground plane, and elevated to some height above ground.

There may be a misconception among some operators that a small ground plane at the base of a vertical whip antenna will make the radiated field spread out into space as it would above an infinite ground plane (radiate zero power at angles below the ground plane). This idea is false. Regardless of the diameter of the ground plane, even if it were 100 wavelengths in diameter, as long as it is hoisted up in the air, and the observer is **beyond** the edge of the ground plane, there will be radiation below the "horizon" plane. The principal cause is "edge diffraction." It causes the angle of maximum radiation to be lifted above the horizon, and

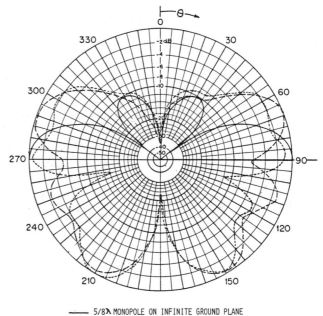

Fig. 4 — Measured vertical patterns of a 5/8-wave whip connected to a length of coaxial line with no ground plane.

strong radiation to be directed **below** the horizon.

For vertical antennas mounted at the center of ground planes several wavelengths in diameter, the loss of radiation intensity toward the horizon is about 6 dB, compared to the same antenna on an infinite ground plane.[8]

For the case of a 5/8-wave monopole mounted on a circular ground plane 1/2 wavelength in diameter, the shape of the radiation pattern is greatly changed compared to the theoretical pattern with an infinite ground plane. Measured patterns of both a 5/8-wave and a 1/4-wave whip mounted on a circular ground plane 1/2 wavelength in diameter are presented in Fig. 5, along with the theoretical pattern of the 5/8-wave whip on an infinite ground plane. There appears to be very little to choose between the 1/4-wave and 5/8-wave whips on the 1/2-wavelength ground plane. Both patterns are very "fat" compared to the theoretical pattern, such that the 3-dB gain factor has been entirely lost.

The substitution of four 1/4-wave radials for the solid metal ground plane can be expected to give substantially the same performance. It might be a little more degraded by the fact that the four radials cannot intercept all of the antenna current flowing out from the outer conductor of the coaxial cable at the feed point. Some of the current still flows along the outside of the outer conductor below the radials, resulting in pattern distortion.

3) An antenna mounted on a hand-held transceiver.

Hand-held transceivers used in the VHF/UHF bands typically use helically loaded flexible whip antennas (rubber duckies) for compactness and convenience. These antennas, however, are notably

103

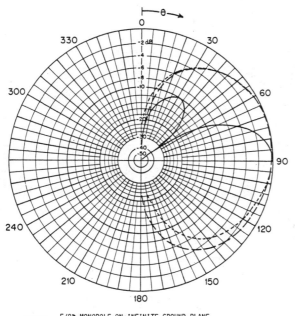

Fig. 5 — Measured vertical patterns of 1/4-wave and 5/8-wave monopoles mounted on a circular ground plane 1/2-wave in diameter. For comparison, the theoretical vertical pattern of 5/8-wave monopole on an infinite conducting plane is shown.

——— 5/8λ MONOPOLE ON INFINITE GROUND PLANE
– – – 5/8λ MONOPOLE ON λ/2 DIAMETER GROUND PLANE
— · — 1/4λ MONOPOLE ON λ/2 DIAMETER GROUND PLANE

quite inefficient. As a result, pull-out antennas of various designs are offered to provide higher radiation efficiency.

The HT and its antenna constitute a dipole. One side consists of the whip antenna; the other includes all the components in the unit which are directly or capacitively connected to the outer conductor of the coaxial connector to which the antenna is attached. At the 2-meter band, the length of this lower half of the dipole is typically less than 0.1 wavelength. The transmitter's role is to place an RF voltage between the two halves of the dipole.

The current distribution along both sides of the dipole is sketched in Fig. 6 for whip lengths of 1/2 and 5/8 wavelengths. Note that the feed point is at a current minimum in the case of the 1/2-wave whip, and the current flowing down onto the lower half of the dipole is consequently low. With the 5/8-wave whip, the current at the feed point is much larger than for the 1/2-wave whip. Furthermore, there is a current phase reversal in the lower part with relation to the phase in the upper portion. The radiation intensity toward the horizon for the case of the 5/8-wave whip should, therefore, be weaker for the same power delivered to the antenna terminals than for the 1/2-wave whip. The stronger excitation of the lower part of the dipole when the 5/8-wave whip is used results in greater coupling of RF energy into the operator's hand and face, compared to when the 1/2-wave whip is used.

Experimentally measured radiation patterns of a 2-meter HT with a helically loaded whip, a 1/2-wave whip, and a 5/8-wave whip are shown superimposed in Fig. 7. The three measured patterns were all made at the same power level. The first pattern measurement

with the helical antenna was repeated at the end in order to ensure that the radiated power had not changed during the series of measurements. This pattern agreed with the first to within 0.1 dB.

Note that the radiation intensity toward the horizon is less for the 5/8-wave whip (about 2 dB) than for the 1/2-wave whip, as predicted.

4) An antenna mounted atop the mast of a sailboat.

This is essentially the same situation as in Case 1. Monopole whips mounted atop sailboat masts are frequently seen. They are principally used in the 156- to 162-MHz marine VHF band. If the coaxial cable comes through a hollow metal mast, as is often the case, the dipole is constructed with the whip on one side and the entire mast on the other, along with various attached stays and shrouds. The radiation pattern (which is difficult to measure) must certainly consist of a mass of thin lobes resulting from radiation from the various conductors which are long compared to the wavelength. This structure of lobes will rock with the boat.

The 5/8-wave whip has a large amount of current at the feed point and performs more poorly than the 1/2-wave whip. The 1/2-wave whip should also be avoided. The proper solution is to use a base-excited vertical whip with an integral, properly designed decoupling system. This guarantees that virtually all antenna currents are confined to the radiator, and little current is excited on the structure below.\9

5) The antenna is mounted on the metal body of an automobile.

In spite of the many years during which VHF and UHF whips have been mounted on vehicles, all of the data regarding the relative performances of the 1/4-, 1/2-, and 5/8-wave whip antennas have not been collected. A car top represents a finite, elevated ground plane; thus the famous 3-dB gain of a 5/8-wave monopole over a 1/4-wave monopole cannot be claimed.

The measured patterns of 1/4-wave and 5/8-wave monopoles over a circular ground plane (Fig. 5) testify to this statement.

A whip antenna mounted on a car can also be regarded as a dipole, one side being the whip, the other the car. Again, continuity of current at the feed point demands that the total current flowing into the base of the whip be exactly equal to the total current flowing out over the car body from the outer conductor of the coaxial feed cable, at the base of the whip. Here, there may be some advantage of a 1/2-wave whip over 5/8 wavelength, because the smaller feed point current of the half wave results in lower current spreading out from the feed point over the car body. As a result, the radiated field from the 1/2-wave whip is probably less sensitive to the mounting location on the vehicle than the field from the 5/8-wave whip. Asymmetry in the mounting location may favor one over the other. For example, trunk-lid mounting on a sedan may favor the 5/8 over the 1/2 wavelength in the forward direction, because the reversed current phase at the base of the 5/8 wave is shielded by the rising car top, in the forward direction. In the reverse direction (with trunk-lid mount) the 1/2 wavelength would probably win.

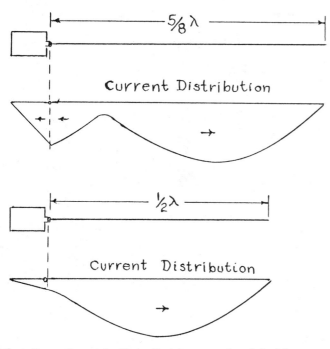

Fig. 6 — Current distributions on hand-held transceivers with 1/2-wave and 5/8-wave whip antennas.

Where 5/8-Wavelength Pays Off

It seems that up to this point, except for its use as a broadcast antenna, the 5/8-wave monopole has little to offer. There is one application, however, for which there is no controversy. This is the 1-1/4-wave center-driven dipole — a dipole in free space in which the length of each leg is 5/8 wavelength. Such an antenna can and does deliver the 3-dB gain figure over a center driven 1/2-wave dipole. This is a well known antenna type, sometimes called the "extended double Zepp." No infinite ground plane is needed to make this antenna perform; the antenna itself has the same geometry and current distribution as a 5/8-wave monopole plus its "image" in an infinite conducting plane.

The 1-1/4-wave center-driven dipole has been used either as a stand-alone antenna, or as an element in an array, driven by a balanced 2-wire open transmission line or a coaxial line with a balun. It may also be used vertically, fed by a coaxial line led up through the lower 5/8-wave element, which must be hollow! Resonant decoupling sleeves must be used in this case to inhibit coupling between the lower radiating element of the dipole and the metallic structure below.\10

Impedance Considerations

So far, nothing has been said about the input impedance of the various lengths of whip antennas. There is reason to suspect that some antenna enthusiasts believe that if an SWR meter reveals no reflected power from an antenna, the antenna must be radiating properly. This assumption is not correct. Lack of reflected power indicates only that the antenna input impedance has been matched to the impedance of the coaxial line. It does not tell how the radiated power is distributed in space. However, even though this article is concerned exclusively with the radiation properties of various antennas, a few words on antenna input impedance are in order.

The input impedance of a 5/8-wave whip above a ground plane 1/2 wavelength or more in diameter has a resistive component close to 50 ohms, and a capacitive reactance that depends considerably on the whip diameter, typically in the range of 50 to 150 ohms. To match the antenna to 50-ohm coaxial cable, an inductor in series with the base should be provided to tune out the capacitive reactance. The simplicity of accomplishing the match is probably one reason for the popularity of the 5/8-wave whip (especially from the manufacturer's standpoint!).

A 1/2-wave whip above a ground plane has an input resistance at its resonant frequency that depends strongly on the whip diameter. Typically this resistance is in the range of 1000 to 2000 ohms. Matching to 50 ohms is readily accomplished with an L-network at the base, which may consist of an inductive reactance shunting the antenna terminals, and a series capacitance to tune out the residual inductive reactance. Practical realizations of such a network can be built in about the same physical volume as the inductor for the 5/8-wave whip.

Measurement Techniques

Free-space radiation-pattern measurements of antennas that eliminate the effects of reflected signals from the earth require expensive experimental facilities. This includes tall, widely spaced towers, and directive antennas. A much simpler method that applies to the vertical pattern of whip antennas designed to radiate omnidirectional, vertically polarized signals, is to support the antenna **horizontally**, about 1 wavelength above the earth on a turntable that is mounted flush with the earth. A test signal is transmitted from the rotating antenna to a fixed, horizontally polarized receiving antenna at a distant point, over level ground. The receiving antenna is mounted at the same height above ground as the rotating antenna. The received signal is amplified and filtered from adjacent-channel interference using a tunable receiver, whose detected output is measured on a chart recorder that is connected by servos to the turntable.

The received signal inevitably contains a directly radiated component and a ground-reflected component. As long as the antenna being measured is rotated in the horizontal plane, and the spacing between the transmitting and receiving antennas is much greater than the heights of both antennas above the earth, the way that the direct and ground-reflected signals phase together at the receiving antenna will be essentially independent of the rotation angle. The measured pattern will then be that of the vertical pattern of the antenna, when it is mounted in its normal, vertical orientation. The validity of this method can be checked experimentally by measuring the radiation pattern of a well constructed reference dipole, whose free-space radiation pattern is accurately known. This procedure was followed, using antenna measurement facilities at the University of Washington.

References

\1 Ballantine, Stuart, "On the Optimum Wavelength for a Vertical Antenna Over Perfect Earth," **Proc. I.R.E.**, Vol. 12, p. 833, December 1924.

\2 See Ref. 1.

\3 Elliott, Robert S., **Antenna Theory and Design**, Prentice Hall, 1981, Chapter 7.

\4 Morita, T., "Current Distributions on Transmitting and Receiving Antennas," **Proc. I.R.E.**, Vol. 38, 1950, pp. 898 to 904.

\5 Stutzman, W. L., and G. A. Thiele, **Antenna Theory and Design,** John Wiley and Sons, 1981, Chapter 5.

\6 See Ref. 1.

\7 Terman, F. E., **Radio Engineer's Handbook,** McGraw Hill, 1943, p. 843.

\8 Balanis, C. A., **Antenna Theory — Analysis and Design,** Harper and Row, 1982, p. 515.

\9 Reynolds, D. K., **Facts About Proper VHF Vertical Antenna Design,** AEA, Inc., P. O. Box 2160, Lynnwood, WA 98036.

\10 See Ref. 9.

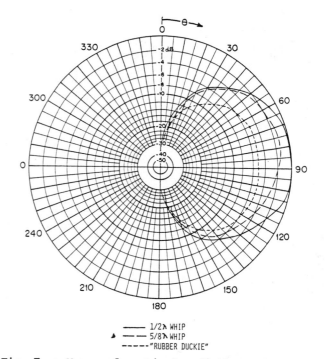

——— 1/2λ WHIP
— — — 5/8λ WHIP
- - - - - "RUBBER DUCKIE"

Fig. 7 — Measured vertical radiation patterns of hand-held transceiver with "rubber duckie," 1/2-, and 5/8-wave antennas.

Antennas of
Reduced Size

Optimum Design of Short Coil-Loaded High-Frequency Mobile Antennas

By Bruce F. Brown,* W6TWW

*1241 Arroyo Seco Dr., Campbell, CA 95008

Optimum design of short high frequency mobile antennas results from a carefully considered balance among loading-coil Q factor, loading-coil position in the antenna, ground loss resistance, and the length-to-diameter ratio of the antenna. To realize this optimum balance, a thorough understanding of how these antenna parameters interact is required. This paper presents a mathematical approach to designing mobile antennas for maximum radiation efficiency, and is an analysis of the parameters affecting optimum placement of a loading coil.

The optimum loading-coil location could be found experimentally, but it would require many hours of design, construction, and measurements. A faster and more reliable way of determining optimum coil location is to write a mathematical program, using a personal computer. With the aid of the computer, all parameters except for one can be held constant, allowing exploration of the effects of a varying parameter. When plotted graphically, the data reveals that the placement of the loading coil is critical if we are to obtain maximum radiation efficiency is to be obtained.

Radiation Resistance

To determine antenna radiation efficiency, power losses caused by resistive losses in the antenna system and radiation loss must be defined. Radiation loss is expressed in terms of radiation resistance. Radiation resistance is defined as the value of resistance that accounts for the portion of power input to the antenna system that is radiated, rather than dissipated as I^2R loss. In the interest of brevity, the variables used in the equations that follow are defined once in the text, and summarized in Table 1 for easy reference.

Radiation resistance of vertical antennas shorter than 45 electrical degrees (1/8 wavelength) is approximately[1]

$$Rr = \frac{h^2}{312} \qquad \text{(Eq. 1)}$$

where Rr is the radiation resistance in ohms, and h is the height of the antenna in electrical degrees.

Antenna height in electrical degrees is expressed by

$$h = \frac{L}{984} \times f(MHz) \times 360 \qquad \text{(Eq. 2)}$$

where L is the antenna height in feet, and f(MHz) is the operating frequency in megahertz. End effect is purposely omitted, to ensure that an antenna is electrically long. Thus, resonance at the design frequency can be obtained easily by removing a turn or two from the loading coil.

Eq. 1 is valid only for antennas having a sinusoidal current profile and no reactive loading. It can be used, however, as a starting place to derive an equation that is useful for short antennas not having a sinusoidal current profile.

From Fig. 1, it can be seen that the current profile on an antenna 90 electrical degrees high (1/4 wavelength) varies as the cosine of the height in electrical degrees. It is also apparent that the current profile of the top 30 degrees of the antenna is essentially linear. It is this linearity that allows derivation of a simpler, more useful equation for radiation resistance.

The radiation resistance of an electrically short base-loaded vertical antenna can be conveniently defined in terms of a geometric figure, a triangle, as shown in Fig. 2.[2] The radiation resistance is

$$Rr = KA^2 \qquad \text{(Eq. 3)}$$

where K is a constant that will be derived shortly, and A is the area of the triangular current profile in degree-amperes.

Degree-amperes is expressed by

$$A = \frac{1}{2} h \times I_{base} \qquad \text{(Eq. 4)}$$

By equating Eqs. 1 and 3 and solving for K, we get

$$K = \frac{h^2}{312 \times A^2} \qquad \text{(Eq. 5)}$$

By substituting the parametric values from Fig. 2 into Eq. 5 we get

$$K = \frac{(30)^2}{312 \times (0.5 \times 30 \times 1)^2} = 0.012820513$$

and by substituting the derived value of K into Eq. 3 we get

$$Rr = 0.012820513 \times A^2 \qquad \text{(Eq. 6)}$$

Eq. 6 is useful for determining the radiation resistance of coil-loaded vertical antennas of 30 degrees or less in height. The derived constant differs slightly from that presented by Laport because of the use of a different equation for

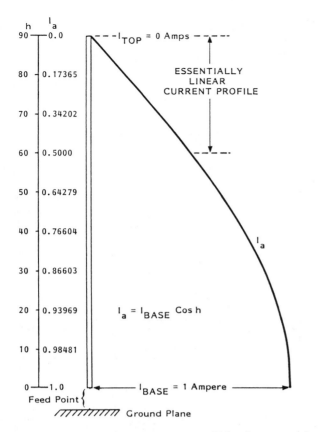

Fig. 1 — Relative current profile in a vertical antenna of height h equal to 90 electrical degrees.

radiation resistance (Eq. 1). I did not round the constant for less decimal places.

When the loading coil is moved up the antenna, the current profile is modified as shown in Fig. 3. The current varies as the cosine of the height in electrical degrees at any point in the base section. Thus, the current flowing into the bottom of the loading coil is less than the current flowing at the base of the antenna.

But what about the current in the top section of the antenna? Well, the loading coil acts as the lumped constant that it is, and maintains the same current flow throughout. Therefore, the current exiting the top of the coil is the same as that entering the bottom of the coil. (This is true for conventional coils. However, radiation from long skinny coils allows coil current to decrease, as in helically wound antennas.) This is easily verified by installing RF ammeters immediately above and below the loading coil in a test antenna. Thus, the coil forces a much higher current into the top section than would flow in the equivalent part of a full 90-degree-high antenna.

The cross-hatched area in Fig. 3 shows the current that would flow in the equivalent part of a 90-degree-high antenna, and reveals that the degree-ampere area of the whip section of the short antenna is greatly increased as a result of the modified current distribution. The extremely high voltage that appears at the top of the loading coil forces this high current flow into the top section. The result is more radiation from the top section than would occur from the equivalent part of a quarter-wave antenna. The current flow in the top section decreases almost linearly to zero at the top. This can also be seen in Fig. 3.

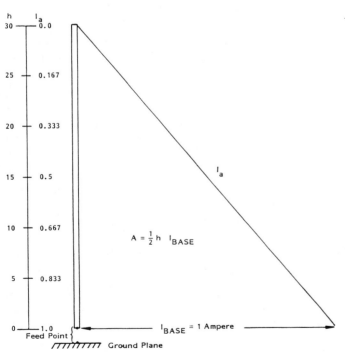

Fig. 2 — Relative current distribution in a base-loaded vertical antenna of height h equal to 30 electrical degrees (linearized). A base loading coil is omitted.

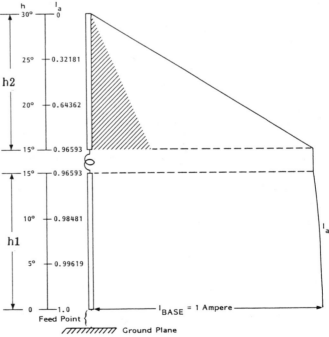

Fig. 3 — Relative current distribution in a center-loaded antenna with base and top sections each equal to 15 electrical degrees in length. The cross-hatched area shows the current profile that would exist in the top 15° of a 90°-high vertical fed with 1 ampere at the base.

109

The degree-ampere area of Fig. 3 is the sum of the triangular area represented by the current profile in the top section, and the nearly trapezoidal current profile in the base section. Radiation from the coil is not included in the degree-ampere area because it is small and difficult to define. Whatever radiation occurs can be considered a bonus.

The degree-ampere area is expressed by

$$A = \frac{1}{2} [h1 \times (1 + \cos h1) + h2 \times (\cos h1)] \qquad \text{(Eq. 7)}$$

where h1 is the electrical height in degrees of the base section, and h2 is the electrical height in degrees of the top section.

The degree-ampere area (calculated by substituting Eq. 7 into Eq. 6) determines the radiation resistance when the loading coil is at any position other than the base of the antenna. Using these equations, radiation resistance has been calculated and plotted against loading-coil positions for 8- and 11-foot antennas. Eight feet is a typical length for commercial antennas, and 11-foot antennas are about the maximum practical length that can be installed on a vehicle. Calculations were made for three different frequencies. With reference to Fig. 4, the curves reveal that the radiation resistance rises almost linearly as the loading coil is moved up the antenna. They also show that the radiation resistance rises rapidly as the frequency is increased. If the analysis were stopped at this point, an erroneous conclusion that the loading coil should be at the top of the antenna could be reached. That is not so, and will be apparent shortly.

Loading Inductance Required

Calculations of the loading-coil inductance needed to resonate a short antenna can be done easily and accurately by using the antenna transmission-line analog described in an article by Boyer.[3] For a base-loaded antenna, Fig. 2, the loading-coil reactance required to resonate the antenna is

$$XL = -j \, Km \times \cot h \qquad \text{(Eq. 8)}$$

where XL is the inductive reactance required, and -j indicates that the antenna presents capacitive reactance at the feed point. (This reactance must be canceled by a loading coil.) Km is the mean characteristic impedance, to be defined in Eq. 9.

The mean characteristic impedance of an antenna is expressed by

$$Km = 60 \times \left[\left(\log_e \frac{2H}{a} \right) - 1 \right] \qquad \text{(Eq. 9)}$$

where H is the height of the antenna (excluding length of the loading coil) in English or metric units, and a is the radius of the antenna in like units of measurement.

From inspection of Eq. 9, it can be seen that decreasing the height-to-diameter ratio of an antenna by increasing the radius results in a decrease in mean characteristic impedance, Km. With reference to Eq. 8, a decrease in Km decreases the inductive reactance required to resonate an antenna. As will be shown later, this will increase radiation efficiency. Of course, in mobile applications we quickly run into wind-loading problems if we attempt to use an antenna with too large a diameter.

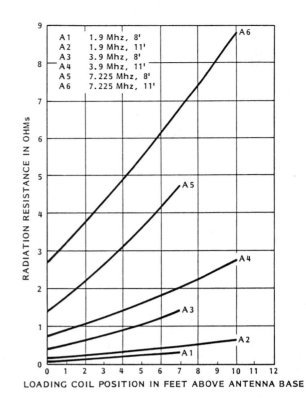

Fig. 4 — Radiation resistance plotted as a function of loading coil position.

If the loading coil is moved from its base position to the antenna midsection, the antenna is divided into a base and top section, as depicted in Fig. 3. The loading-coil inductive reactance required to resonate the antenna when the coil is above the base position is expressed by

$$XL = j \, Km2 \times (\cot h2) - j \, Km1 \times (\tan h1) \qquad \text{(Eq. 10)}$$

In mobile-antenna design and construction, the top section of a whip usually has a much smaller diameter than the base section. Because of this, it is necessary to compute separate values of Km for the top and base sections. Km1 and Km2 are the mean characteristic impedances of the base and top sections, respectively.

Loading-coil reactance curves for the 75-meter antennas of Fig. 4 have been calculated and plotted in Fig. 5 to show the influence of the loading-coil position in the antenna on the reactance required for resonance. The curves in Fig. 5 show that the required reactance decreases with longer antennas. Significantly, they reveal that the required loading-coil reactance grows at an increasingly rapid rate after the coil passes the center of the antenna. Since the highest possible loading-coil Q factor is needed, and since optimum Q is attained when the loading-coil diameter is twice the loading coil length, the coil would grow like a smoke ring above the center of the antenna, quickly reaching an intolerable size.[4] It is for this reason that in all my computations the highest loading-coil position is limited to one foot from the top of the antenna.

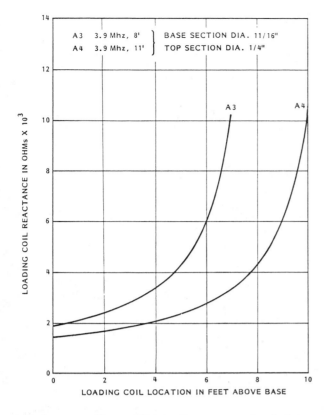

Fig. 5 — Loading coil reactance required for resonance, plotted as a function of coil height above the antenna base. The resonant frequency is 3900 kHz.

Loading Coil-Resistance

Loading-coil resistance is one of the resistive losses that consumes power which could otherwise be radiated. Heat loss in the loading coil is not of any benefit, so it should be minimized by using the highest Q possible. Loading-coil loss resistance is a function of the coil Q factor and is expressed by

$$R_c = \frac{X_L}{Q} \qquad \text{(Eq. 11)}$$

where R_c is the loading-coil loss resistance, X_L is the loading-coil inductive reactance, and Q is the coil figure of merit.

From inspection of Eq. 11, we can see that, for a specific value of inductive reactance, the loss resistance will be lower for higher-Q coils.

High Q can be obtained by going to diameter coils having a diameter-to-length ratio of two, by using larger diameter wire, by using more spacing between turns, and by using low-loss polystyrene supporting and enclosure materials. Loading-coil turns should not be shorted for tuning purposes because shorted turns degrade Q. The highest possible Q is needed. Pruning to resonance should be performed by removing turns from the coil instead.

Radiation Efficiency

The ratio of power radiated to power fed to an antenna determines the radiation efficiency. It is expressed by

$$E = \frac{Pr}{Pi} \times 100 \text{ percent} \qquad \text{(Eq. 12)}$$

where E is radiation efficiency in percent, Pr is the power radiated, and Pi is the power fed to the antenna at the feed point.

In a short coil-loaded mobile antenna, a large portion of the power fed to the antenna is dissipated in ground and coil resistances. A relatively insignificant amount of power is also dissipated in the antenna conductor resistance and in leakage resistance of the base insulator. Because these latter two losses are both very small and difficult to define, they are neglected in calculating radiation efficiency.

Another loss, worthy of note in passing, is coupling-network loss. Since we are concerned only with power fed to the antenna in the determination of radiation efficiency, coupling-network loss is not considered in any of the equations. Suffice it to say that coupling networks should be designed for minimum loss in order to maximize the transmitter power available to the antenna.

The radiation efficiency equation may be rewritten and expanded as follows:

$$E = \frac{I^2 \times Rr \times 100}{I^2 \times Rr + I^2 \times Rg + (I \cos hl)^2 \times Rc}$$

$$\text{(Eq. 13)}$$

where I is the antenna base current, Rg is the ground loss resistance, and Rc is the coil loss resistance.

Each term of Eq. 13 represents the power dissipated in its associated resistance. By inspection it can be seen that all the I values cancel out, simplifying the equation to

$$E = \frac{Rr}{Rr + Rg + Rc \times \cos^2 hl} \times 100$$

$$\text{(Eq. 14)}$$

For base-loaded antennas the term $\cos^2 hl$ reduces to unity and may be omitted.

Ground Loss

Eq. 14 shows that the total resistive losses in the antenna system are

$$Rt = Rr + Rg + Rc \times \cos^2 hl \qquad \text{(Eq. 15)}$$

where Rt is the total resistive loss. Ground loss resistance can be determined by rearranging Eq. 15 as follows:

$$Rg = Rt - Rr - Rc \times \cos^2 hl \qquad \text{(Eq. 16)}$$

Rt may be measured in a test antenna installation on a vehicle using an antenna noise bridge or an RX bridge. Rr and Rc can be calculated.

Ground loss is a function of vehicle size, placement of the antenna on the vehicle, and conductivity of the ground over which the vehicle is traveling. Only the first two variables can be

111

controlled. While I don't discourage anyone from going to a small vehicle to reduce fuel consumption, be aware that larger vehicles provide a better ground plane than smaller ones. In any event, the vehicle ground plane is only partial, so the result is considerable RF current flow (and ground loss) in the ground around and under the vehicle.\5

By raising the antenna base as high as possible on the vehicle, the ground losses are decreased. This results from a decrease in antenna capacitance to ground, which increases the capacitive reactance to ground. This, in turn, reduces ground currents and ground losses.

This effect has been verified by installing the same antenna at three different locations on two different vehicles, and by determining the ground loss from Eq. 16. In the first test, the antenna was mounted 6 inches below the top of my large station wagon, just behind the left rear window. This placed the antenna base 4 feet 2 inches above the roadway, and resulted in a measured ground loss resistance of 2.5 ohms. The second test used the same antenna mounted on the left rear fender of a mid-sized sedan, just to the left of the trunk lid. In this test, the measured ground loss resistance was 4 ohms. The third test used the same mid-sized sedan, but with the antenna mounted on the rear bumper. In this last test, the measured ground loss resistance was 6 ohms.

So we have the same antenna seeing three different ground loss resistances purely as a result of the antenna mounting location and size of the vehicle. It is important to note that the measured ground loss increased as the antenna base was nearer the ground. The importance of minimizing ground losses in mobile antenna installations will become apparent shortly.

Efficiency Curves

With the previously defined equations, a computer program was written in BASIC and used to calculate the radiation-efficiency curves depicted in Figs. 6 through 9. These curves were calculated for 75- and 40-meter antennas of 8- and 11-foot lengths. Several values of loading-coil Q factor were used, for both 2 and 10 ohms of ground loss resistance. For the calculations, the base section is half-inch-diameter electrical metallic tubing (EMT), which has an outside diameter of 11/16 inch. The top section is fiberglass bicycle-whip material covered with Belden braid. These are readily available materials which can be used by the average amateur to construct an inexpensive but rugged antenna.

Upon inspection, these radiation-efficiency curves reveal some significant information: (1) higher coil Q produces higher radiation efficiencies; (2) longer antennas produce higher radiation efficiencies; (3) higher frequencies produce higher radiation efficiencies; (4) lower ground loss resistances produce higher radiation efficiencies; (5) higher ground loss resistances force the loading coil above the antenna center to reach a crest in the radiation-efficiency curve; and (6) higher coil Q sharpens the radiation-efficiency curves, resulting in the coil position being more critical for optimum radiation efficiency.

The reason that the radiation-efficiency curves reach a crest and then begin to decline as the loading coil is raised farther is because of the rapid

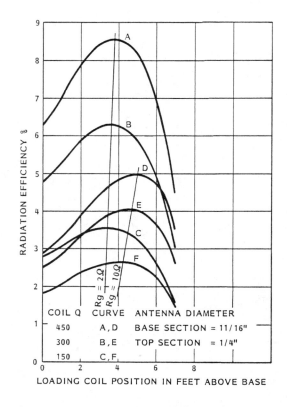

Fig. 6 -- Radiation efficiency of 8-ft antennas at 3900 kHz.

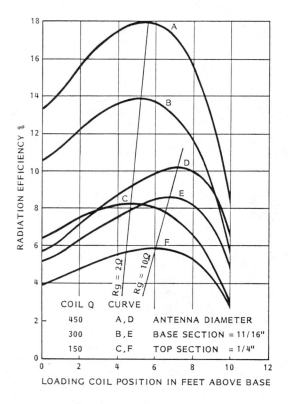

Fig. 7 -- Radiation efficiency of 11-ft antennas at 3900 kHz.

increase in loading-coil reactance required above the antenna center. Refer to Fig. 5. The rapid increase in coil size required for resonance results

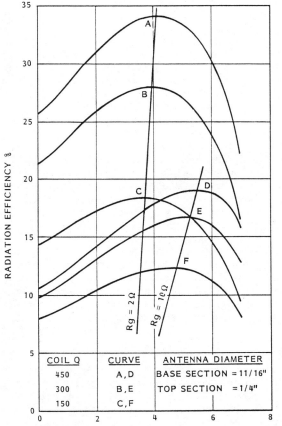

RADIATION EFFICIENCY %

COIL Q	CURVE	ANTENNA DIAMETER
450	A, D	BASE SECTION = 11/16"
300	B, E	TOP SECTION = 1/4"
150	C, F	

LOADING COIL POSITION IN FEET ABOVE THE BASE

Fig. 8 — Radiation efficiency of 8-ft antennas at 7225 kHz.

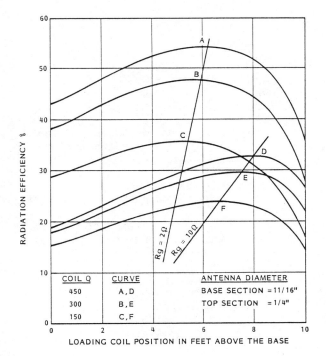

RADIATION EFFICIENCY %

COIL Q	CURVE	ANTENNA DIAMETER
450	A, D	BASE SECTION = 11/16"
300	B, E	TOP SECTION = 1/4"
150	C, F	

LOADING COIL POSITION IN FEET ABOVE THE BASE

Fig. 9 — Radiation efficiency of 11-ft antennas at 7225 kHz.

in the coil loss resistance increasing much more rapidly than the radiation resistance. See Fig. 4. This results in decreasing radiation efficiency.

A slight reverse curvature in the curves between the base-loaded position and the one-foot coil-height position is visible. This results from a shift in the curve because of insertion of a base section of larger diameter than the whip when the coil moves above the base.

The curves in Figs. 6 through 9 were calculated with constant, but not equal-diameter base and whip sections. Because of wind loading, it is not desirable to increase the diameter of the whip section. However, the base-section diameter can be increased within reason to further improve radiation efficiency. Fig. 10 was calculated for base-section diameters ranging from 11/16 of an inch to three inches. The curves reveal that a small increase in radiation efficiency results from larger diameter base sections.

The curves in Figs. 6 through 9 show that radiation efficiencies can be quite low in the 75-meter band compared to the 40-meter band. They would be lower still in the 160-meter band. To gain some perspective on what these low efficiencies mean in terms of signal strength, Fig. 11 was calculated using the following equation:

$$dB = \log \frac{100}{E} \qquad \text{(Eq. 17)}$$

where dB is the signal loss in decibels, and E is efficiency in percent.

The curve in Fig. 11 reveals that an antenna having 25% efficiency has a signal loss of 6 dB (one S unit) below a quarter-wave vertical reference antenna operated over perfect ground. An antenna efficiency in the neighborhood of 6% will produce a signal strength on the order of two S units or 12 dB below a quarter wave reference vertical. By careful optimization of mobile-antenna design, it can be seen that signal strengths from mobiles can be fairly competitive with those from fixed stations of comparable power.

Impedance Matching

The input impedance of short high-Q coil-loaded antennas is low. As an example, an optimized 8-foot antenna for 3,900 kHz having a coil Q of 300 and a ground loss resistance of two ohms has a base input impedance of about 13 ohms. This low value of impedance causes a standing wave ratio in 50-ohm coaxial line of 3.85:1 at resonance. This high SWR is not compatible with the low SWR requirements of solid-state finals. In addition, the bandwidth of short vertical antennas is very narrow. This severely limits the capability to maintain transmitter loading with a small frequency change.

Impedance matching can be accomplished by means of L networks or impedance-matching transformers, but the narrow bandwidth limitation remains. A more elegant solution to the impedance matching and narrow bandwidth problem is to install an automatic tuner at the antenna base.\6 Such a device matches the antenna and coaxial line properly, and permits operation over a wide frequency range.

Conclusion

The mathematical modeling technique employing a personal computer revealed that loading-coil Q factor and ground loss resistance greatly influence the optimum loading-coil position in a short vertical

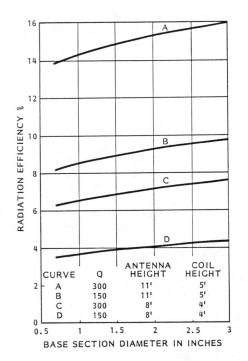

Fig. 10 — Radiation efficiency plotted as a function of base section diameter.
Frequency — 3900 kHz.
Ground loss resistance — 2 ohms.
Whip section — 1/4-in diameter.

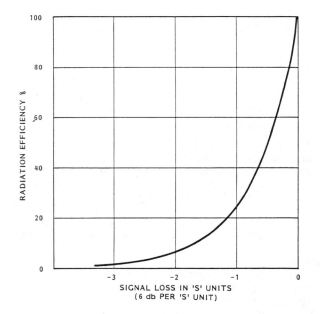

Fig. 11 — Mobile-antenna signal loss as a function of radiation efficiency, compared to a quarter-wave vertical antenna over perfect ground.

antenna. It also shows that longer antennas, higher coil Q, and higher operating frequencies produce higher radiation efficiencies.

The tools are now available to tailor a mobile antenna design to produce maximum radiation efficiency.\7 A missing ingredient is the ready availability of very high-Q commercial coils. It is hoped that this paper will stimulate both the demand for

and the production of such coils.

End effect has not been included in any of the equations to assure that the loading coil will be slightly larger than actually required. Antenna pruning to resonance should be done by removing coil turns, rather than by shorting turns or shortening the whip section. Shortening the whip section reduces radiation efficiency, by both shortening the antenna and moving the optimum coil position. Shorting turns degrades the coil Q factor.

Shortened Dipoles

The mathematical modeling technique can be applied to shortened dipoles by using zero ground loss resistance and by doubling the computed values of radiation resistance and feed-point impedance. However, radiation efficiency does not double. The reason is that to complete the other half of the shortened dipole requires a second loading coil. The doubling of radiation resistance is offset by the coil resistance of two coils, leaving the radiation efficiency unchanged. There is a gain in radiation efficiency over a vertical antenna worked against ground because the dipole configuration permits ground loss resistance to be zeroed out of the model.

Acknowledgements

I wish to acknowledge and express my appreciation to Rudy Severns, N6LF, for patiently urging me to prepare this paper, his numerous suggestions, experimental verification of the current flow into and out of the loading coil, and assistance in proofreading the manuscript. I also wish to express my appreciation to Syd Padrick, WB6HXJ, for his assistance in making the ground loss measurements, and to John McCollum, WB6LVD, for his word-processing assistance.

Table I

A = area in degree-amperes
a = antenna radius in English or metric units
dB = signal loss in decibels
E = efficiency in percent
f(MHz) = frequency in megahertz
H = height in English or metric units
h = height in electrical degrees
h1 = height of base section in electrical degrees
h2 = height of top section in electrical degrees
I = I_{base} = 1 ampere base current
K = 0.012820513
Km = mean characteristic impedance
Km1 = mean characteristic impedance of base section
Km2 = mean characteristic impedance of top section
L = length or height of the antenna in feet
Pi = power fed to the antenna
Pr = power radiated
Q = coil figure of merit
Rc = coil loss resistance in ohms
Rg = ground loss resistance in ohms
Rr = radiation resistance in ohms
XL = loading-coil inductive reactance

References

\1 Smith, Carl E. and Earl M. Johnson, "Performance of Short Antennas," **Proceedings of the IRE,** October 1947.

\2 Laport, Edmund A., **Radio Antenna Engineering,** 1952, p. 23.

\3 Boyer, Joseph M., "Antenna-transmission line analog," **Ham Radio,** May 1977.

\4 Terman, Frederick E., **Radio Engineering Handbook,** Third Edition, 1943, p. 74.

\5 Belrose, J. S., "Short Antennas for Mobile Operation," **QST,** Sept. 1953.

\6 Brown, Bruce F., "Tennamatic: An Auto-Tuning Mobile Antenna System", **73,** July 1979.

\7 A computer program is available from B Square Enterprises, P.O. Box 71, Campbell, CA 95008.

Short Loaded Half-Wave Dipole Design — The Easy Way

By H. L. Ley, Jr.,* N3CDR

*Herbert L. Ley Associates, Inc., P. O. Box 2047, Rockville, MD 20852

An important but often neglected element in a station is the antenna. The present-day ham operates at a disadvantage with respect to his colleagues of 25 to 50 years ago. Available real estate for antenna installation is elusive, and antennas are now being placed in limited quarters such as an apartment or boat. **The ARRL Handbook** and **The ARRL Antenna Book** offer useful, but limited information on problems like these. In the past, I've inserted a loading inductor at an appropriate point in the antenna. By measuring the resonant frequency with a grid-dip meter or noise bridge, I would prune the coil until I brought the shortened, loaded antenna to resonance.

Enter, the Home Computer

With the availability of home computers, it is now possible to approach short loaded half-wave dipole design in a rational manner. The basic formulas have been known for some time, but they are complex and long combinations of calculations are sometimes required to arrive at a single solution. Most hams are unwilling to spend laborious hours evolving the best design for a given physical layout.

I have brought the necessary equations together into a cassette tape computer program for both the Texas Instruments 99/4A home computer and the Sinclair ZX-81/Timex-Sinclair 1000 home computer with 16 K added memory. The programs are convenient to use and permit anyone to explore more than a half-dozen options for a short-loaded dipole in a given physical layout during a 15-30 minute period. If sufficient demand develops, I will also provide the program in 5 inch single-sided, single-density or double-density disk format for the Heath/Zenith 89 computer in Microsoft BASIC running on a CPM operating system. Persons desiring full documentation of the program are invited to write me directly.

Basics of Loaded Dipole Design

Before I provide examples of the programs, the basic factors involved in the design of loaded dipoles should be reviewed. This will help us to understand how the computer programs work. Fig. 1A presents the voltage and current distributions in a conventional, full-length center-fed half-wave dipole. Current is maximum at the center feed point where the voltage is lowest. Conversely, voltage is maximum and current is zero at the ends of the antenna.

Figure 1B presents the same curves for a resonant dipole loaded in the center of each element. It is half the length of the antenna shown in the first view and its voltage and current in the first and last half of each element is about the same as in Fig. 1A. At the location of the loading coils, there is an abrupt change in both voltage and current distribution along the antenna. The loading coils replace the missing portions of the original, full length antenna, and carry a significant voltage and current drop over the length of the coils.

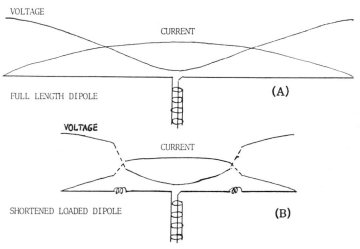

Fig. 1 — A comparison of currents and voltages in a full length and shortened loaded dipole.

The short half-wavelength loaded dipole exhibits two distinctive features in comparison with a full length antenna. The harmonic resonant frequencies cannot be predicted as easily as they can in a full-length dipole.\1 For this reason, short loaded dipoles are not recommended for harmonic operation. If this antenna is used for more than one frequency, arrangements must be made to short out unused inductors or have a separate plug-in loading inductor for each band of operation.

The second difference is antenna bandwidth, which is decreased in comparison with a conventional dipole.\2 Careful adjustment of the loading-coil values used to center the resonant frequency at the desired operating frequency is required if you are to obtain a low SWR on the feed line.

If any antenna is resonant (exhibits only resistance and no reactance at the resonant frequency), it is possible to calculate the voltage at the feed point by the conventional power equations:

$$E\,\mathrm{rms} = \sqrt{P \times R}$$

$$I\,\mathrm{rms} = \sqrt{\frac{P}{R}}$$

where,

P = power transmitted to antenna (1:1 SWR) in watts.

R = radiation resistance of antenna in ohms (as measured with an impedance or noise bridge).

At a power level of 100 W and a radiation resistance of 50 ohms, the appropriate values are E = 70.7 V and I = 1.4 A. These values are important because they indicate the insulation and current-carrying capacity required at the feed point. Maximum radiation occurs in the high central current portions. This establishes rule no. 1 of a loaded antenna design: Make the distance from the feed point to the loading coil as long as possible, consistent with your space requirements.

Given a short antenna of a fixed length, the inductance of the loading coils required to bring it to resonance increases in value as they are moved from the center of the antenna toward the ends. This is important when working with loading coils. Indeed, at the very tips of the short antenna, a loading coil would require **infinite** inductance to resonate it. Rule no. 2 therefore states: Place the loading coil at a position in a short antenna such that it does not require an unreasonable value of inductance to resonate the antenna.

Because rules 1 and 2 conflict with each other, a compromise is necessary. This can best be achieved by developing a series of options for possible antennas to fit a given space in terms of the distance of the loading coils from the feed point, and the inductance values required in the various positions to resonate the antenna.

As a general rule, placing the loading coils at the feed point should be avoided for two reasons. First, the high-current portion of the antenna that is responsible for most radiation is replaced by a coil. Second, the high current at that location results in the highest I^2R heat losses in the inductor. The best position then is somewhere along the antenna element away from the feed point. This location depends on physical constraints and the size of the loading coil you have.

Previous Articles on Loaded Dipole Design

The basic mathematics for computation of the inductance required for loading short dipoles to resonance was developed by J. Hall, K1TD.[3] His article was acquired by Sander who wrote an abbreviated computer program in BASIC for the Apple II Plus Computer, with Orr and Rasor later revising the program for the TRS-80 computer.[4,5] Both programs are highly abbreviated and do not include subprograms for calculating the matching impedance at the feed point by the method of Gooch et al,[6] or for calculating the winding data for both the loading and matching inductances. The consolidated program I wrote includes all of these features. To illustrate the use of the program, an example of an actual design is presented.

Design Examples

Problem 1: Design a short-loaded dipole, 16 ft long, to fit a balcony space in an apartment.

After loading the cassette into the computer, you will receive a prompt for the operating frequency. I picked 7.125 MHz, the center of the 40-meter novice band (see Fig. 2). I specified a total length of 16 ft, and a distance of 2 ft from the center for the loading coils. My next step was to decide if I would use wire or aluminum tubing for the antenna. For the first run, I chose no. 18 wire. The computer calculated a value of 33.0 µH for each loading coil. Those are **big** coils! What would have been the value if I chose Reynolds (TM) 0.75-in diameter with 0.047-in wall thickness tubing? It takes only a minute to obtain the new results.

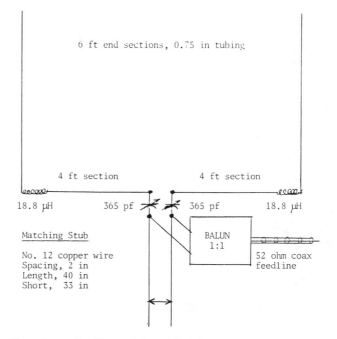

Fig. 2 -- Configuration of indoor loaded dipole for 7.10-7.15 MHz.

```
ENTER OPERATING FREQ IN MHZ = 7.125
ENTER TOTAL LENGTH IN FEET = 16
ENTER DISTANCE FROM CENTER TO LOADING
   COIL IN FEET = 2
DO YOU WANT TO USE THE WIRE TABLE? (Y OR N)
ENTER WIRE GAUGE (10 TO 30) = 18
LOADING COIL IS 33.000237 UHY
DO YOU WISH TO CALCULATE WINDING DATA FOR
   COIL? (Y OR N)
```

Computer printout of calculation for 16 ft no. 18 wire antenna.

When tubing is used for the antenna, a much smaller inductor is needed for resonance (see Fig. 3). The calculated value is 19.5 µH — a considerable decrease in inductance and coil size.

```
ENTER OPERATING FREQ IN MHZ = 7.125
ENTER TOTAL LENGTH IN FEET = 16
ENTER DISTANCE FROM CENTER TO LOADING COIL
    IN FEET = 2
DO YOU WANT TO USE THE WIRE TABLE? (Y OR N)
WHAT IS ELEMENT DIAMETER IN DECIMAL INCHES
    = 0.75
LOADING COIL IS 19.469348 UHY
DO YOU WISH TO CALCULATE WINDING DATA FOR COIL?
    (Y OR N)

INPUT COIL DIAMETER, DECIMAL INCHES = 2
INPUT COIL LENGTH, DECIMAL INCHES = 4
NUMBER OF TURNS REQUIRED = 30.886859
DO YOU WISH TO CALCULATE ANOTHER VERSION OF
    COIL? (Y OR N)
```

Fig. 3 — Computer printout of calculation for 16 ft 0.75-in tubing antenna.

The computer program next prompted me to decide if I wanted to calculate the winding data for the inductance. If I answered "Y" for yes, the computer would perform the necessary mathematics, giving an answer of 31 turns of wire, 2 in diameter and 4 in long. The ARRL L/C/F Slide Rule Calculator may used as an alternative. It is easier to use if you're working with commercial coil stock where turns per inch is the critical parameter. In that case, answer "N" for no to the computer query.

The program provides an opportunity to calculate the center matching inductance, connected to the feed point, to match a specified feed line impedance (see Fig. 4). The program calculates the antenna impedance (unless you have a measured value to substitute), the matching inductive reactance and the required inductance at the operating frequency. A word of caution is appropriate here. If the antenna length originally chosen is longer than a half wavelength at some frequency, the program will return a negative value for the inductance. The necessary matching element is then capacitive, and its value can be calculated by hand from the reactance at chosen the frequency.\7

These calculations can be continued for the same 16-ft antenna, varying the distance from the center for the loading coils, and for other bands. Results from such calculations are shown in Table 1. This data clearly shows the increase in value of the loading coils as they are moved away from the feed point. It also shows the rising antenna impedance as frequency is increased.

```
ANTENNA IMPEDANCE IS ESTIMATED BY MULTIPLYING
    52 OHMS BY RATIO OF TOTAL LENGTH TO FULL
    DIPOLE LENGTH = 12.664889 OHMS
DO YOU HAVE A MEASURED VALUE TO SUBSTITUTE?
    (Y OR NO)
INDUCTIVE REACTANCE (XL) = 29.506238 OHMS
MATCHING INDUCTANCE = 0.65918917 UHY
THIS COIL WILL HAVE TO BE TRIMMED TO LOWEST SWR
DO YOU WISH TO CALCULATE WINDING DATA FOR
    COIL? (Y OR N)
```

Fig. 4 — Computer printout of calculation for center matching inductor.

Problem 2: Design a short, loaded dipole, 20 ft long, to fit on the ceiling of a room measuring 8 by 10 ft.

In this problem, the antenna is to be installed indoors suspended from the ceiling of a room. The second part of Table 1 gives the results of a series of calculations, for 0.75-in aluminum tubing, based on inserting the loading inductances 2, 3 and 4 ft from the feed point for a resonant frequency of 7.125 MHz. The most desirable physical arrangement for this antenna turns out to be a "U" shape with the loading coils placed 4 ft from the feed point on an 8 ft side. Each element is then extended with a 6-ft length of tubing at right angles to the 8-ft center portion. With this configuration, a loading coil of 18.78 µH should bring the antenna to resonance for 7.125 MHz. For 14.2 and 21.15 MHz, the appropriate values are 3.29 and 0.43 µH.

This computer-designed antenna was fabricated and installed in a room on the fourth floor of a multi-unit apartment building. It was suspended 10 in below the ceiling from hooks installed in the ceiling. Nylon cord was used as insulators. A Plexiglas corner fitting held the tubing sections at right angles and provided a mounting for the loading coils installed at the corners. When initially checked for resonance with a Palomar noise bridge and a general-coverage receiver, the antenna was found to be resonant approximately 10% below the design frequencies. Because of capacitive loading by house wiring and heating ductwork, the antenna impedance was estimated to be about 15 ohms at 7.125 MHz and 40 ohms at 21.15 MHz. Although the loading coils could have been pruned to bring the antenna to resonance, it was elected to use a simpler method — series capacitive correction.

The configuration worked out for 40 m is shown in Fig. 2. This design permits the resonance adjustment to be made at the antenna feed point. The receiver and noise bridge are connected to the antenna with a half wavelength of coax cable to determine the proper settings of the series capacitors.

When tuned to resonance with the matching stub shown, the antenna impedance at 7.125 MHz was found to be approximately 50 ohms. At one point, the capacitors were placed in series with each of the loading coils at the corners of the antenna on the feed-point side. The electrical characteristics of the antenna appeared identical to those it exhibited with the capacitors at the center, but the inconvenience of setting two capacitors located at opposite ends of the room was bothersome. Thereafter, the feed point location was used! To simplify the adjustment further, the two variable capacitors were mounted on a piece of Plexiglas and ganged by three gears. One was placed on each capacitor shaft and the third, an idler gear, was mounted between the two capacitors. A tuning knob was attached to the shaft of the idler gear to permit one-hand tuning.

The logic behind this method of tuning the antenna to resonance is simple. The antenna appears to be longer than it actually is when loaded near building structures. It therefore exhibits inductive reactance. By adding an equal capacitive reactance to cancel the inductive reactance, the antenna is adjusted to resonance.

Using this antenna over a period of one month on the 7.10-7.15 MHz band, I confirmed contacts with both New York and Wisconsin using a Heath HW-8 QRP transceiver. Maine, Massachusetts, New York, Pennsylvania, Ohio, Indiana, Illinois, Wisconsin and West Virginia were contacted with a 30-year-old crystal-controlled Johnson Adventurer transmitter (30 W measured output) and a Collins 51J-4 receiver. Other contacts with New Hampshire and Vermont have not yet been confirmed. The antenna favored the north and west, probably because the center section was hung in a general east-west direction. Little effort was put into 21-MHz contacts, other than local. I had to prove to my satisfaction that it could be adjusted to resonance on that band as well. Most of those I contacted were surprised, to put it mildly, to learn that I was working them on a 20-ft, indoor loaded dipole!

Myth or Reality?

When the 16-ft loaded tubing dipole shown in Table 1 was tested in a balcony installation, the antenna resonated at a lower frequency than 7.125 MHz for which it was designed. Inductor trimming was necessary and this observation caused doubts about the validity of the equations used for the design work. It must be stated that the antenna was located parallel to and about 3 ft distant from both a rain gutter and a steel balcony railing. These obstacles could be expected to load the antenna capacitively, thereby lowering its resonant frequency and its efficiency as a radiator.

A 1:1 balun was installed at the feed point of the same antenna. With the aid of KA3FLA, the loading coils were trimmed for 7.125 MHz in the balcony installation. The antenna was erected in open space about 7 ft above moist ground. A Palomar R-X noise bridge was connected to the antenna with 50 ft of RG-58 foam coax cable. The cable's measured velocity factor was 0.770, making the cable one-half electrical wavelength at 7.57 MHz. Several additional lengths of cable were available for coupling to the 50-ft length to make total cable lengths corresponding to half wavelengths at 7.050, 7.125, 7.200 and 7.300 MHz. A Heath HD-1250 dip meter was used as a radiating extrapolation oscillator coupled with a small pickup coil to a Heath IM-4100 frequency counter. It gave an accurate frequency measurement of the noise null in a general-coverage SWL-type receiver to determine the resonant frequency of the antenna with the noise bridge. Much to my surprise, the resonant frequency of the antenna installed in an open area was about 7.5 MHz. A series of four frequency measurements of the resonant frequency of the antenna was made with the 50-ft coax section alone as the best compromise length. These were averaged to a value of 7.47 MHz. The radiation resistance of the antenna was estimated with the use of resistance standards to be 15 ohms (the resistance scale of the Palomar noise bridge is not suitable for accurate measurements).

When the resonant frequency of 7.47 MHz was entered into the computer program, a calculated value for the loading coil placed 2 ft from the feed point was 17.59 μH. The ARRL L/C/F Slide Rule Calculator yielded a calculated value for the coil installed in the antenna of 17.7 μH. The two values differed by only 0.6%, validating rather well the equations used in the computer program.

No effort was made to accurately measure the resonant frequency of the antenna with smaller loading coils pruned to produce resonance in the 20- and 15-meter bands in the balcony installation because of the limited choices of length of the coax feed line. We were fortunate that the 50-ft length of coax approximated in its electrical length a half wavelength at the resonant frequency of the antenna erected in open space.

This experience proved to me that the formulas used in the computer program were adequate to predict values of loading coils used to resonate the short antenna. It also provided evidence that the detuning effects of parallel conductors in proximity to the antenna could be major, as much as 5% in my case.

Matching the Antenna and Feed line

The 40-meter, short loaded dipole presents a major problem of matching to a 52-ohm coax feed line. The radiation resistance of the 16-ft antenna approximates 13 ohms. Delrin rod (0.625-in diameter) is recommended for the center joint of the antenna, for spacers for the break at the loading coils, and for the end insulators. It possesses excellent electrical, mechanical and machining characteristics. If the 0.75-in aluminum tubing is slit for approximately 1.5 in at the ends, a no. 10 bolt through the tubing and Delrin rod compresses the tubing around the rod to make a stable joint. The bolt, preferably aluminum to minimize corrosion problems, should be placed about 0.75 in from the end of the tubing.\8 This matching problem has been approached with a gamma match in many beam antennas, and a similar approach could be used with the short loaded dipole. However, one reason for the popularity of the gamma match in beams is that it permits the use of an unbroken radiator, giving greater structural strength to the antenna. I have not found that to be a necessary feature in apartment antennas, so I have resorted to an older matching network called the "hairpin match," described by Gooch.\9 The "hairpin" is installed at the feed point between two electrically isolated halves of the antenna. If made of no. 8 copper wire spaced 2 in apart, it can be adjusted easily by means of a shorting bar that can be tightened on the wires to provide good electrical contact. A useful approximation of the inductance of such a hairpin is provided by the following formula:

$$L \text{ (μH)} = \frac{\text{Length to shorting bar, in}}{26}$$

To calculate the approximate length of the hairpin to the shorting bar, the formula can be rearranged, as follows:

$$\text{Length to shorting bar, in} = L(\text{μH}) \times 26$$

These relationships are approximately correct for lengths of 6 to 24 in. The graph in Gooch's article is not a linear function, so these formulas are not exact. Nevertheless, because the shorting bar has to be adjusted empirically for lowest SWR, they do provide a starting point for adjustments. The computer provides the value of the inductance required at the feed point to convert the low impedance of the loaded dipole to any specified feed line impedance. As with the loading coils, if the computer returns the value of a negative inductance, the matching element is capacitive, not inductive. Obviously, the matching element could be a typical helically wound coil rather than the hairpin. The hairpin is so much easier to adjust that I have found it preferable to a conventional coil.

Summary

The availability of the home computer permits rapid, reasonably accurate calculations of a number of options for installation of shortened, loaded dipoles in limited space. My experience with the use of my program developed from Amateur Radio related literature has been convincing with its predictability of the resonant frequency of the short loaded dipole in open space. It is also clear from the results of installation of apartment antennas that the detuning effect of conductors in the vicinity of the antenna is not easily predictable. In brief, the computer program will provide a starting point for an antenna design in a dwelling area, but it does not substitute for the "cut and try" pruning method used to compensate for the detuning effects of the environment on the antenna.

I would be interested to correspond with other amateurs using the basic formulas employed in my computer program for antenna design. Please enclose an s.a.s.e. for a prompt reply. Good luck, and good radiation!

References

\1 National Bureau of Standards, Circular C74, **Radio Instruments and Measurements** (Washington: U.S. Government Printing Office, 1937), pp. 75-78.

\2 ARRL Staff, **The ARRL Handbook**, 55th Ed., (Newington: ARRL, 1978), p. 608.

\3 J. Hall, "Off-Center-Loaded Dipole Antennas", **QST**, Sept. 1974, p. 28.

\4 D. Sander, "A Computer Designed Loaded Dipole Antenna", **CQ**, Dec. 1981, p. 44.

\5 B. Orr, "Ham Radio Techniques", **Ham Radio**, Apr. 1983, p. 52.

\6 J. D. Gooch, O. E. Gardner and G. L. Roberts, "The Hairpin Match", **QST**, Apr. 1962, p. 11.

\7 Formula for calculating capacitance from capacitive reactance:

$$C(F) = \frac{1}{2 \times pi \times f\,(H) \times Xc\,(ohms)}$$

\8 Delrin rod may be purchased from plastic suppliers or from Read Plastics, Inc., 12331 Wilkins Ave., Rockville, MD 20852, tel. (301) 881-7900.

\9 See Ref. 6.

Table 1

Antenna Design Examples

Total Length Ft/Matl	Freq. MHz	Load Coil Ft From Center	Load Coil Induct µH	Antenna Imped Ohms	Match React Ohms	Match Induct µH
16/#18 wire	7.125	2	33.0	12.7	29.5	0.66
16/0.75 tubing	7.125 " "	2 3 4	19.46 22.36 26.49	12.7 " "	29.5 " "	0.66 " "
16/0.75 tubing	14.2 21.15	2 2	3.95 1.10	25.2 37.6	50.5 84.0	0.57 0.63
20/0.75 tubing	7.125 " "	2 3 4	15.02 16.65 18.78	15.8 " "	34.4 " "	0.77 " "
20/0.75 tubing	14.2 21.15	4 4	3.29 0.43	31.6 47.0	64.6 159.4	0.72 1.20
16/#18 wire	7.125	2	33.0	12.7	29.5	0.66
16/0.75 tubing	7.125 " "	2 3 4	19.46 22.36 26.49	12.7 " "	29.5 " "	0.66 " "
16/0.75 tubing	14.2 21.15	2 2	3.95 1.10	25.2 37.6	50.5 84.0	0.57 0.63
20/0.75 tubing	7.125 " "	2 3 4	15.02 16.65 18.78	15.8 " "	34.4 " "	0.77 " "
20/0.75 tubing	14.2 21.15	4 4	3.29 0.43	31.6 47.0	64.6 159.4	0.72 1.20

Computer program for ZX-81/Times-Sinclair 1000
computer +16k RAM

```
1000 CLEAR
1010 CLS
1020 PRINT "PROGRAM TO DESIGN LOADED DIPOLE"
1030 PRINT "C/R 1983, HERBERT L. LEY ASSOC., INC."
1040 PRINT
1050 PRINT "ENTER OPERATING FREQ IN MHZ"
1060 INPUT F
1070 PRINT "= ";F
1080 PRINT
1090 PRINT "ENTER TOTAL LENGTH IN FEET ";
1100 INPUT A
1110 PRINT "= ";A
1120 PRINT
1130 PRINT "ENTER DISTANCE FROM CENTER TO LOADING COIL IN FEET";
1140 INPUT B
1150 PRINT "= ";B
1160 PRINT
1170 LET X=((234/F)-B)
1180 LET Y=((A/2)-B)
1190 PRINT "DO YOU WANT TO USE THE WIRE TABLE? (Y OR N)"
1200 INPUT Y$
1210 IF Y$="Y" THEN GOSUB 1520
1220 IF Y$="N" THEN GOSUB 1680
1230 LET S1=1E6/(68*PI**2*F**2)
1240 LET S2=1/X*(LN (24*X/D)-1)
1250 LET S3=(1-F*B/234)**2-1
1260 LET S4=1/Y*(LN (24*Y/D)-1)
1270 LET S5=(Y*F/234)**2-1
1280 LET L=S1*(S2*S3-S4*S5)
1290 PRINT
1300 PRINT "LOADING COIL IS ";
1310 PRINT L;" UHY"
1320 PRINT
1330 PRINT "DO YOU WISH TO CALCULATE WINDING DATA FOR COIL? (Y OR N)"
1340 INPUT Y$
1350 IF Y$="Y" THEN GOSUB 2090
1360 CLS
1370 PRINT
1380 PRINT "DO YOU WISH TO CALCULATE VALUE OF CENTER MATCHING COIL? (Y
OR N)"
1390 INPUT Y$
1400 CLS
1410 IF Y$="Y" THEN GOSUB 1730
1420 CLS
1430 PRINT
1440 PRINT "DO YOU WISH TO CALCULATE ANOTHER ANTENNA? (Y OR N)"
1450 INPUT Y$
1460 IF Y$="Y" THEN GOTO 1000
1470 PRINT
1480 PRINT "PROGRAM TERMINATED, TO RUN AGAIN ENTER ""R"""
1490 INPUT R$
1500 IF R$="R" THEN GOTO 1000
1510 STOP
1520 PRINT
1530 PRINT "ENTER WIRE GAUGE (10 TO 30)";
1540 INPUT D
1550 PRINT "= ";D
1560 IF D=10 THEN LET D=.101
1570 IF D=12 THEN LET D=.081
1580 IF D=14 THEN LET D=.064
1590 IF D=16 THEN LET D=.051
1600 IF D=18 THEN LET D=.040
1610 IF D=20 THEN LET D=.032
1620 IF D=22 THEN LET D=.025
1630 IF D=24 THEN LET D=.020
1640 IF D=26 THEN LET D=.016
1650 IF D=28 THEN LET D=.013
```

```
1660 IF D=30 THEN LET D=.010
1670 RETURN
1680 PRINT
1690 PRINT "WHAT IS ELEMENT DIAMETER IN DECIMAL INCHES";
1700 INPUT D
1710 PRINT "= ";D
1720 RETURN
1730 CLS
1740 PRINT "ENTER FEED LINE IMPEDANCE IN OHMS";
1750 INPUT ZT
1760 PRINT "= ";ZT
1770 PRINT
1780 PRINT "ANTENNA IMPEDANCE IS ESTIMATED BY MULTIPLYING 52 OHMS BY
RATIO OF TOTAL LENGTH TO FULL DIPOLE LENGTH ";
1790 LET ZA=52*(A/(468/F))
1800 PRINT "= ";ZA;" OHMS"
1810 PRINT
1820 PRINT "DO YOU HAVE A MEASURED VALUE TO SUBSTITUTE? (Y OR N)"
1830 INPUT Y$
1840 IF Y$="N" THEN GOTO 1880
1850 PRINT "INPUT VALUE IN OHMS";
1860 INPUT ZA
1870 PRINT "= ";ZA
1880 LET XL=ZT*(SQR (ZA/(ZT-ZA)))
1890 PRINT "INDUCTIVE REACTANCE (XL) = ";
1900 PRINT XL;" OHMS"
1910 PRINT
1920 LET L=XL/(2*PI*F)
1930 PRINT "MATCHING INDUCTANCE = ";
1940 PRINT L;" UHY"
1950 PRINT
1960 PRINT "THIS COIL WILL HAVE TO BE TRIMMED TO LOWEST SWR"
1970 PRINT
1980 PRINT "DO YOU WISH TO CALCULATE WINDING DATA FOR COIL? (Y OR N)"
1990 INPUT Y$
2000 IF Y$="Y" THEN GOSUB 2090
2010 RETURN
2020 PRINT "DO YOU WANT TO CALCULATE ANOTHER ANTENNA? (Y OR N)"
2030 INPUT Y$
2040 IF Y$="Y" THEN GOTO 1000
2050 PRINT
2060 PRINT "PROGRAM TERMINATED, TO RUN ENTER ""R"""
2070 INPUT R$
2080 IF R$="R" THEN GOTO 1000
2090 CLS
2100 PRINT "INPUT COIL DIAMETER, DECIMAL INCHES ";
2110 INPUT D
2120 PRINT "= ";D
2130 PRINT
2140 PRINT "INPUT COIL LENGTH, DECIMAL INCHES";
2150 INPUT B
2160 PRINT "= ";B
2170 LET R=D/2
2180 LET N=SQR ((L*(9*R+10*B))/R**2)
2190 PRINT
2200 PRINT "NUMBER OF TURNS REQUIRED = ";
2210 PRINT N
2220 PRINT
2230 PRINT "DO YOU WISH TO CALCULATE ANOTHER VERSION OF COIL? (Y OR
N)"
2240 INPUT Y$
2250 IF Y$="Y" THEN GOTO 2090
2260 RETURN
```

Miscellaneous Antennas

Dielectric Antennas for the 10-GHz and Higher Amateur Bands

By David Andersen,* KK9W

*3240 McCormick Rd., W. Lafayette, IN 47906

Dielectric antennas are useful devices for coupling power from metal waveguides into free space. They are used primarily in the microwave and millimeter regions of the radio frequency spectrum, an area where the dimensions of familiar wire antennas become too small for practical implementation. This paper presents a qualitative description of the electric field configuration in the antenna, and provides a set of design rules for implementing dielectric antennas.

The rectangular waveguide is a popular transmission line in the microwave region of the RF spectrum. In most systems, the guide is operated in what is known as the TE_{10} mode. TE stands for transverse electric, and means that there is no component of the electric field in the direction of propagation in the waveguide. The only component of the electric field present is that which points from one long side to the other long side of the rectangle for cross section. In addition, the field undergoes a half sinusoidal variation as an object moves along the long side of the cross section. At one edge, it starts at zero, moves through a maximum, and back to zero at the other edge. This type of field variation is specified by the subscript 10 in the mode name.

To efficiently radiate energy from a rectangular waveguide, a system must be used to match the impedance between the guide and free space. A horn antenna, constructed from conductive material, is one such method used. The cross section of the horn becomes progressively larger as the wave propagates through it. Finally, the horn is terminated, after the cross section is sufficiently large that the effects of the metal boundaries are minimized.

The major drawback of using only a horn antenna is that it does not exhibit high-gain characteristics. As a result, it does not have a narrow beamwidth. This problem can be overcome in several ways. The use of a horn and a second reflector, such as a parabolic dish, can be employed to increase the gain of the horn antenna. The horn illuminates the parabolic dish and reradiates, or reflects, the energy with a much narrower beamwidth and higher gain. A second method involves the use of a dielectric antenna.

Dielectric material allows radio waves to propagate through it without loss. It also decreases the free space velocity of the radio wave from $c = 3 \times 10^8$ meters per second. The insulating material between capacitor plates, the plastic material used in coaxial transmission lines, and the material between the traces of a microstrip transmission line and its groundplane are examples of dielectrics.

Dielectric material is characterized by a number called the relative dielectric constant. This is a unitless number that indicates the amount a radio wave is slowed by the dielectric medium. The relative dielectric constant is denoted E_r and is defined as:

$$Er = \left[\frac{c}{V}\right]^2$$

where V is the velocity of the radio wave in the dielectric medium. A number sometimes used instead of the relative dielectric constant is called the index of refraction. It is denoted n and is simply the square root of the relative dielectric constant.

The dielectric antenna is known as a leaky-wave antenna. It allows some radiation to leak into free space, as the input or feed energy propagates along the structure. The amount and phase shift of the energy which is leaked or radiated can be adjusted by varying the dimensions of the antenna. If the antenna is designed correctly, the input signal, by the time it reaches the end of the antenna structure, should have been completely radiated. Adjusting the phase of the energy radiated also adjusts the sidelobe levels and beamwidth.

There are three general configurations for dielectric antennas. These are specified as the E-plane taper, the H-plane taper, and a combination of these two, the EH taper. As their names suggest, the antennas contain tapers in the respective polarization planes. Experimental work has been done to demonstrate that all of the advantages of dielectric antennas can be realized by using an E-plane taper exclusively.\1 Thus, the type of antenna discussed here will be the E-plane tapered configuration shown in Fig. 1.

The antenna of Fig. 1 consists of several subparts. The first is the metal horn. This can be thought of as the launching mechanism for the radio wave, which is then focused by the dielectric material. Thus, by exciting the dielectric antenna, the horn plays a role analogous to its role in exciting the parabolic dish antenna.

The metal horn excites the dielectric antenna, but the excitation is not 100% efficient. This efficiency can be increased by adjusting the feed taper, denoted L_f in Fig. 1. The active part of the antenna, which does most of the radiating, is the part denoted L_r. This part of the antenna must be long enough to allow a large amount of the excitation energy to be reradiated; however, it must be short enough that construction is possible. Finally, the terminal taper, denoted L_e, is used to terminate any small amount of energy not radiated so that the energy does not reflect off the end of the structure and degrade the radiation pattern.

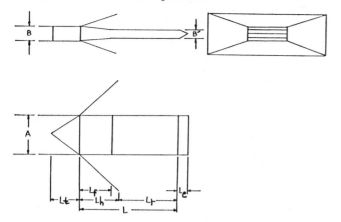

Fig. 1 — Configuration of E-plane tapered antenna.

A qualitative description of the fields inside the dielectric can be obtained by assuming that the fields do not change radically for the first few wavelengths of propagation in the dielectric. The TE_{10} mode of the metal waveguide has fields perpendicular to the direction of propagation. As the wave propagates along the dielectric rod, a component of the electric field parallel to the surface of the antenna must exist in order for the antenna to radiate. This results in a slight curving of the electric field as it propagates. The qualitative changes the electric field undergoes are shown in Fig. 2.

Fig. 2 — Change in E-field along dielectric antenna.

Design principles for leaky-wave antennas have been established by Zucker, and some experimental work has been performed by Kobayashi, et.al., to demonstrate the feasibility of applying the design principles to dielectric antennas.[2] The result of this work has been a set of design rules for dielectric antennas. It is the goal of this paper to demonstrate the use of these design rules for the 10-GHz amateur band.

Several things must be specified in order to begin the design: the frequency of operation, the dimensions of the metal waveguide to be used, and the dielectric constant of the dielectric material. Typical numbers for amateur use might be the following: an X-band frequency of 10.250 GHz, a dielectric constant $E_r = 2.25$, and metal waveguide of type WR-90, which has dimensions of approximately A = 2.5 cm and B = 1.25 cm. Additionally, the free-space wavelength at 10.250 GHz is 2.927 cm.

The next step in the design is to obtain the ratio of the wavelength in free space to the wavelength at the metal guide/dielectric interface. This ratio, denoted Ro, is equivalent to the square root of the effective dielectric constant and can be obtained from propagation curves of the E_{11} mode of

the dielectric waveguide. These plots are available in Ref. 3. For our particular problem, the plot used was Fig. 21 of Ref. 3, and the value obtained for Ro was 1.167.

The next step involves selecting the beamwidth and gain of the desired antenna. Zucker has plotted two experimental curves (see Ref. 2, Figs. 16-10 and 16-11) that can be used. Initially, select the length of the antenna, L, which gives the desired beamwidth and/or gain from Fig. 16-11. For our problem, the value of L selected was L = 17.56 cm or 6 wavelengths. This results in a gain of approximately 18 dB and a beamwidth of 21 degrees. Gain can be increased by increasing L.

When L has been selected, the designer must then determine the average ratio of the wavelength in free space to the wavelength in the dielectric medium along the length of the dielectric rod. This is obtained by using Fig. 16-10 of Ref. 2 and is found, for our example, to be Ra = 1.05.

Now that the numbers Ro, Ra, and L have been specified, the length of the feed taper can be calculated. This is done using the following formulas:

where:

$$L_f = \frac{1}{2}\left[C1 - \sqrt{C1^2 - 4C2} \right]$$

and:

$$C1 = \frac{\lambda_o + 6(Ro + Ra - 2)L}{6(Ro - 1)}$$

$$C2 = \frac{\lambda_o L}{3(Ro - 1)}$$

Additionally, the effective dielectric constant at the end of the feed taper can be calculated as:

$$Rr = \frac{\lambda_o}{3 L_f} - Ro + 2$$

For our example problem, these numbers are:

$$C1 = 25.74 \ cm$$
$$C2 = 102.6 \ cm^2$$
$$L_f = 4.93 \ cm$$
$$Rr = 1.031$$

Using the value of Rr, and referring back to the propagation curves of Ref. 3, the width of the radiating section of the dielectric antenna, B', can be obtained. Here, an approximation is made by assuming the propagation curve for the A/B = 2 situation still applies. The ratio A/B' is actually greater than 2. This is a relatively good approximation, however, because of the weak coupling of the fields to the dielectric at this point. For the example problem, B' = 0.7078 cm.

The last two items to be specified are the horn dimensions, L_h, and the terminal taper, L_e. Experiment has shown that $L_h = 1.1 \ L_f$ and $L_e =$ one-half wavelength are good values to use.

Some of the considerations involved in designing dielectric antennas have been discussed. In addition, a design for an 18-dB gain, 21 degree beamwidth antenna at X-band was formulated. The design rules presented are also applicable to the amateur bands above 10 GHz. I hope that this information is of use to amateurs and will promote greater use of this unique antenna.

References

\1 Kobayashi, S., Mittra, R., Lampe, R., **IEEE Trans. Ant. and Prop.**, V. AP-30, no. 1, p. 54 (1982).

\2 Zucker, F. J., **Antenna Engineering Handbook**, H. Jasik, Ed., New York: McGraw-Hill, 1961, chap. 16.

\3 Goell, J. E., **Bell System Technical Journal**, V. 48, p. 2133 (1969).

\4 Kiely, D. G., **Dielectric Aerials**, London: Methuen and Co. (1953).

\5 Watson, R. B., and Horton Jr., C. W., **Applied Physics**, V. 19, no. 7, p. 661 (1948).

A Crossed-Loop/Goniometer DF Antenna for 160 Meters

Charles F. W. Anderson,* N4KF

*1716 Reppard Rd., Orlando, FL 32803

What would low-band enthusiasts say to a DF antenna with performance comparable to W1FB's 1.5-meter square loop? \1 It measures only one meter on a side and its pattern is rotatable by nothing more than a hand-operated control right on your desk. If this interests you, read on!

The crossed-loop/goniometer direction finder dates back to the early days of wireless. Developed by two Italian engineers, Bellini and Tosi, it consists of two fixed loops, arranged at 90-degree angles. The outputs of these loops are routed through transmission lines leading to two sets of coils. Called primary coils, they are also at right angles to each other. Inside these coils, or between the halves of each, another coil, called the secondary coil, is rotated. The output of the secondary, or pickup coil, feeds the receiver input. These inductors form the goniometer. Fig. 1 shows a typical setup. It is based on descriptions and diagrams in Terman and Henney's radio engineering handbooks, and from Chapter 10 of **Electronic Countermeasures**.\2,3,4 W2IMB also describes a crossed-loop/goniometer in **The Low and Medium Frequency Radio Scrapbook**.\5

Preliminaries

Refer to the Bill of Materials. Most of the items listed, other than the coaxial cable and shielded, twisted-pair line, should be available at any building supply store.

First, cut the pipe using information from the following table. Chamfer the inside edge of each end of all the pieces.

Identification	Length	Number
A pieces	35 in	8
B pieces	25 in	4
C pieces	1 1/2 in	16
Stanchion	54 in	1

Two of the 10-ft pieces of pipe will make six of the **A** pieces, plus the **C** pieces. A third piece of the pipe should be cut to make two more **A** pieces and two of the **B** pieces. The remaining piece of pipe should have the stanchion cut from it, plus the other two **B** pieces.

Cut two 40-ft pieces of the RG-174 coaxial cable. At the exact center of each of these, remove about 3/8-inch of the jacket and outer conductor, using great care not to cut into the dielectric. Wrap the areas with insulating tape.

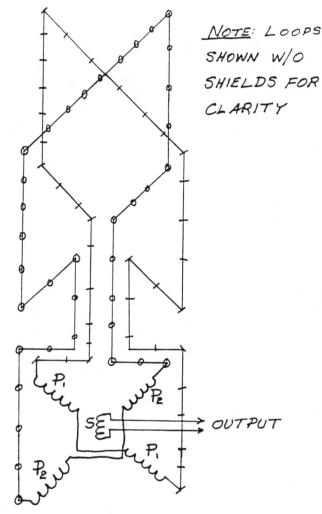

NOTE: LOOPS SHOWN W/O SHIELDS FOR CLARITY

Fig. 1 -- Crossed-loop/goniometer.

Loop Housing Assembly

The housing for the loops is about one meter on a side and stands 1.4 meters high. Refer to Fig. 2. If you follow the steps below with reasonable care, you should have no difficulty in constructing the housing.

At the exact center of each of the cross fittings, cut a 3/4-inch diameter hole on each side. Of the methods I tried, the carpenter's brace-and-bit seemed to work best. Save at least one of the circular pieces you've removed for use later. About 1/2 inch from each of the cross ends, drill a small hole into which you can thread a self-tapping screw. Comparable holes at each port of the Ts and the 45-degree elbows must also be drilled.

Slide one of the cross fittings onto the stanchion piece and position it exactly at the middle. Hold the fitting in place with a wrapping of tape on each side. Three-eighths inch from each end, wrap several turns of tape around the stanchion. These wrappings will keep the top and bottom cross fittings from sliding too far onto the stanchion while the rest of the housing is assembled.

Insert the four **B** pieces into the center cross-fitting ports and hold them in place with self-tapping screws. Now put the other two cross fittings on the ends of the stanchion. Hold them in place by running a length of monofilament line through the stanchion and tying the ends together. At this point, you'll need to make an arrangement to hold the framework in a vertical position while completing the assembly. I suspended mine from a ceiling air duct in the shack, using heavy cord.

Mate a **C** piece with each of the 45-degree elbows, and secure with set screws. Insert one of these into each of the ports of the bottom fitting. Align them accurately, and hold in place with set screws.

Insert four of the **A** pieces into the other side of the elbows, and then put an elbow/C-piece on the ends of each of the **A** pieces. Mate each of the Ts with the four **B** and **C** pieces. Recheck alignment, loosening and tightening the set screws as necessary.

Put four additional elbow/C-pieces on the upper ports of the Ts and the remaining four on the top cross fitting. Now place the last four **A** pieces in position, make final alignment adjustments to get the sides at right angles to each other, then tighten the set screws just enough to keep everything together. This completes the assembly of the loop housing.

If you have room to mount the loop assembly indoors, (attic or air space, for example), you could eliminate the **A** pieces, the **C** pieces, and the 45-degree elbows. This would simplify assembly and reduce the cost. Note my remarks on this later in the article.

Installing the Loops

Spray some silicone lubricant into each of the four sides of the housing. This will help you through the following steps!

Squeeze a split shot onto the end of a piece of monofilament line about 15 ft long. Start the shot down one side of the housing and get it down to the bottom cross fitting by shaking, bumping or any other means of persuasion you might think of. See Fig. 3A.

When the end of the monofilament exits the bottom of the housing assembly, lash the end of one piece of coaxial cable along the monofilament at the top of the assembly. Use a short piece of monofilament for this purpose. See Fig. 3B.

Start pulling on the end of the monofilament to bring the cable down through the housing. There might be a little trouble when the end approaches the first elbow. A little jiggling of the monofilament and cable should get it around the corner. If you run into trouble here, loosen some of the set screws and fish the coaxial-cable end through corner.

Pull the monofilament until the end of the coaxial cable comes out at the bottom. Untie the

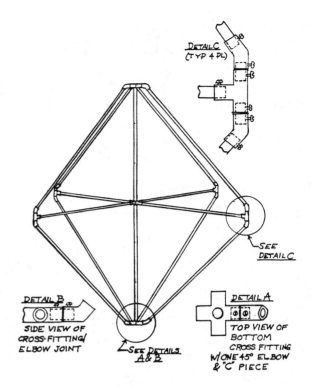

Fig. 2 — Loop housing.

cable from the monofilament, and secure the upper end of the monofilament so that it won't follow along with the cable during the next step (Fig. 3C). The cable should be pulled through the housing until the center is just inside of the top cross fitting. This is shown in Fig 3D.

Repeat these steps using another piece of monofilament and the other end of the coaxial cable. Care should be taken to keep the kinks out of the cable while you're pulling it into the top fitting. The middle of the cable should be centered in the top fitting, with the two ends coming out the bottom.

Attach one of the cable ends to the monofilament which is in the opposite side of the housing. See Fig. 3E. Pull the monofilament out of the top cross fitting until the end emerges, then remove the lashing, and pull the cable through. At this point, be careful about kinks in the cable at the low cross fitting (Fig. 3F).

Attach the other end of the cable to the bottom of the monofilament which is in the side of the housing opposite to the one you've just pulled through, and repeat the previous step. You should now have both ends of the cable at the top of the housing, with monofilament lines in both sides with which to get the cable back to the bottom.

One end of the cable should be connected to the monofilament in the correct side of the housing. Pull it down and out of the bottom cross fitting. Repeat this step with the other end of the cable.

Tie the ends of the coaxial cable out of the way for the time being. Remove the monofilament line used to retain the top and bottom cross fittings. Repeat this procedure from the start for the other length of cable.

Close the hole in the top fitting with the circular piece you saved (step one of the housing assembly) and seal it with silicone adhesive. Loosen each of the set screws slightly and put a bead of silicone around each joint to make a good rain-tight seal. Push the joint back together and retighten the set screw. Assembly of the crossed loops is now complete.

Interfacing the Loops to the Goniometer

The loop outputs in my setup are run directly to the primary coils. Resonating the loops and primary coils with a variable capacitor in series gives higher input levels at the receiver. The addition of preamps at the loop outputs should give even better results. See W1FB's article in April 1984 **QST** for information on preamps.\6 Whichever route you decide to take, an enclosure is needed at the base of the loop assembly in which to make connections to the transmission lines, and to house preamps, if used. Selection of the proper Minibox should be based on what it will contain. For connecting the loop outputs to the transmission lines, I mounted three standoff insulators where each of the sets of loop conductors and transmission lines are connected together. One of each group is used as the tie point for the two ends of the loop shield (the coaxial cable outer conductor) and the shield of the transmission line. The other two are used to connect the inner conductors of the loop to the twisted pair of the associated line. Trim the coax ends as necessary to make these connections.

Mounting the Crossed-Loop Assembly

My loop assembly is currently secured in a tree outside the shack. Reroofing of the house is imminent, which accounts for the unconventional "roost." Later, the assembly will be clamped to a 20-ft PVC pipe (Schedule 80, 2-inch dia.), to be supported at the eaves by a mast bracket. You might want to consider roof-mounting the loop assembly. One approach is to use a TV mast base to support the bottom, and substitute cross fittings for the Ts at the side corners. These could be used in combination with 90-degree elbows to provide support. The four legs made of Ts or elbows should be of the right length for the pitch of the roof and could rest on the surface. Some kind of tie-down should be provided. Adequate guying, using nonconducting lines, might preclude the need for the legs.

Transmission Lines

If the distance from the location of the loop to your operating table doesn't exceed 20 ft, almost any kind of shielded, twisted pair should be adequate. If the distance is greater, I suggest using one of the RF-grade lines. I'm using Raychem 10602, which is designed for the 1-megabit data bus used in military airborne installations. It has a nominal loss of 1.4 dB per 100 ft at 1 MHz. Gore and others make similar cables. RG-22 or RG-108 would also be suitable. The two transmission lines are routed into my shack using grommeted holes in the Plexiglas panel that has replaced the conventional windowlight and serves as the feedthrough for antenna cables. From there, they are run to the chassis in which the goniometer is installed.

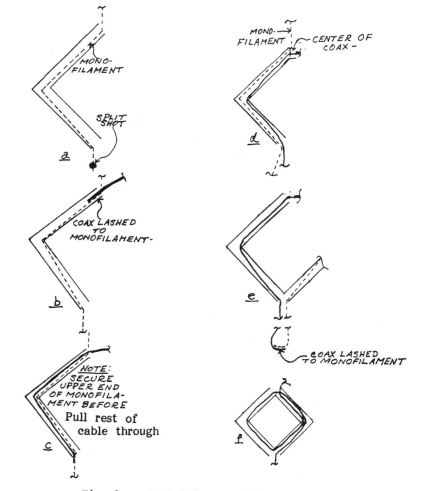

Fig. 3 -- Installing coaxial cable.

Goniometer

This part of the setup consists of the primary and the secondary coils. A good method for fabricating this was the subject of much head-scratching. Part of the answer was found on the household-items shelf of a store. A plastic container, with cover, measuring about 9 x 9 x 3 inches, proved to be the right form for the primary coils. It included room for the rotatable pickup coil, as shown in Fig. 4.

Preparing the plastic box for winding the coils requires a little work. A raised lip at the top of the box mates with the groove in the lid. Mark the exact center of the lip and of the ridge that runs around the outside of the box on each side. At each of the marked spots on the lip, remove 1 1/4 inches of the lip (5/8 inch on each side of the center marks) down to the level of the ridge.

Now, starting 3/16 inch from the center of each side of the ridge, cut a shallow, flat-bottomed notch 1/8-inch wide. These notches (two on each side of the box) provide winding spaces for the primary coils. On two adjacent edges of the box, just outside of the notches, drill holes to mount the standoff insulators. These will be the terminals for the primary windings and the transmission lines. This is also shown in Fig. 4. Mark the exact center of the bottom of the box, and drill a hole to install the shaft bushing.

To prevent the sides of the box from being pulled inwards by the primary windings, a block of styrofoam about the same thickness as the depth of the box is cut and shaped to fit snuggly inside. Here's a tip to spare yourself some extra work. Cut the styrofoam into eight pieces, corner-to-corner and side-to-side. It will be easy to remove these pieces after the windings are in place.

After the bushing was installed, I had another problem. How do I space the windings away from the bushing and above it? My wife provided the solution in the form of a champagne cork, which she found in one of the kitchen drawers. A little work with an X-acto™ knife converted the cork into what was needed. Remove the shank of the cork so that about 1/8 inch is left. Drill a 3/8-inch hole through the center of the cork. This spacer holds the windings away from the shaft. For the primary windings, solder the end of a no. 26 enamel-insulated wire to one of the standoffs. Wind six turns around the box, using the notches to hold the wire, cross over on the top side of the box, and wind six more turns. Cut the wire, making certain that the windings are neither too loose nor so tight that the lid won't fit. Solder the end of the wire to the other standoff. Repeat these instructions for the other primary windings. Place the spacer over the shaft (bottom side of box), spread the windings over it to clear the shaft, and remove the styofoam. The coils measured about 39 µH.

Finding a suitable form for the pickup coil also presented a problem. Here again, styrofoam was employed. The coil was wound on a block measuring 5-1/4 x 1-3/4 x 1-1/2 inches. It was calculated that the coil should have an inductance of about 50 µH (reactance of about 600 ohms at 1.8 MHz) to match the balanced input of the ancient SX-110 receiver used for 160 meters. To keep the number of turns within reason and to improve the Q, a broadcast-band ferrite loop core, 5-1/8 x 5/16-inch diameter was inserted into a hole cut lengthwise through the block. The hole was offset to allow the end of the coil-rotating shaft to be fastened in place. The coil was wound with no. 26 enamel-insulated wire; 16 turns spaced about the diameter of the wire on each side of the block, with a space between the halves of the winding sufficient for clearance for the

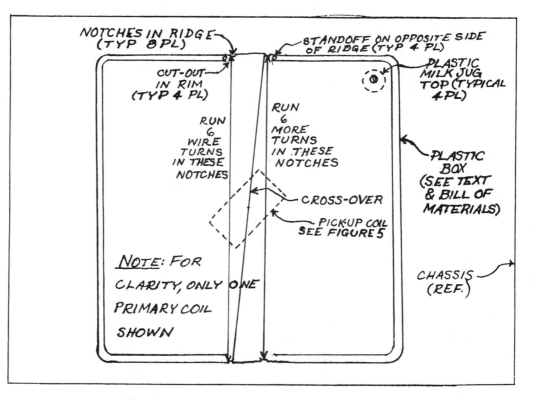

Fig. 4 — Goniometer box with one primary coil.

shaft. A 1/4-inch hole was drilled part way through the block and the shaft was inserted and bonded with silicone adhesive. See Fig. 5.

To make the output line for the secondary coil, cover a length of fine twisted pair with shielding braid. The twisted pair I used came from a defunct in-the-ear headphone unit. The shielding braid used was removed from a piece of subminiature coaxial cable. Connect the twisted pair to the secondary coil leads. Then make one turn or more of the output line around the coil shaft so that it can be rotated easily. Connect the shield to the chassis at both ends. The goniometer assembly is now ready to be mounted. The four milk jug tops are used to space it away from the chassis. Install the plastic lids after the box has been secured in place.

At a convenient point on the chassis, two standoffs are mounted. The twisted-pair leads are terminated at this spot. The shield should be connected to the chassis near the standoffs. Another length of the shielded transmission line makes the connections from this point to the receiver. Also at this time, the loop output lines should be connected to the primary coils. The shields of these lines are connected to the chassis close to the coil terminals.

The shield should be bonded to the chassis at all entry and exit points to minimize stray pickup of RFI and BC station harmonics. The chassis bottom plate is part of this overall shielding implementation. Install a knob on the pickup coil shaft, and the goniometer is complete.

Receiver Input Connections

As mentioned above, my receiver has a balanced 600-ohm input. Connecting the goniometer output line only involved attaching the twisted-pair leads to the antenna terminals and the shield to the ground terminal. Coaxial cable was tried earlier for the connection from the pickup coil to the receiver, but didn't work well in my case. If your receiver has a 50-ohm input, consider using a balun.

Operation

There is nothing complicated about using the setup. Rotate the pickup coil to null out noise or QRM.

Some DF Dope and Test Results

DF antenna patterns are affected by **any** conducting object in the vicinity. While flight-testing radio installations during WW II, I spent many hours making calibrations of what is called "quadrantal error." For a medium-sized aircraft (wingspan of approximately 50 ft), this error can run from about 15 degrees at 200 kHz to over 20 degrees in the BC band (it goes higher as the frequency increases).

Quandrantal error curves usually show little or no deviation at the 0-, 90-, 180- and 270-degree points relative to the station being received. However, the error increases to levels of the order mentioned above at or near the intercardinal points. In addition, at those points, the nulls of the pattern will not be as sharp. The major factor contributing to the deviations is the "static dipole" field (also known as the induction electric field), a component of the complex near-field geometry in the vicinity of any antenna. This field varies inversely as the cube of the distance from radiating or reradiating objects, and is therefore the predominant field in the immediate vicinity. Also, its intrinsic impedance is quite high, compared with the 377 ohms of the far field. To minimize some of the effects of this field, loops are provided with electrostatic shields; hence, the RG-59 which Doug DeMaw used for his big loop, the RG-58 of his "W1FB 4T-ES" loop and the RG-174 in the unit described herein. Don't expect that your crossed-loop/goniometer setup will give accurate bearings on local stations or interference sources. The wiring, piping, ducting, metal flashing and rain gutters and spouts of your house will all contribute to errors, so think twice before installing your loop indoors!

To demonstrate the importance of loop shielding, I made some checks. Using a lash-up version of the W1FB 1.5-meter square loop, I ran a test on a local station near the high end of the BC band. With the loop shield ends not connected to the shield of the line running to the receiver input, hardly any change in S-meter reading was seen as the loop was rotated through 180 degrees. With the loop shield connected properly, a sharp null of at least 30 dB was observed. Similar results were obtained

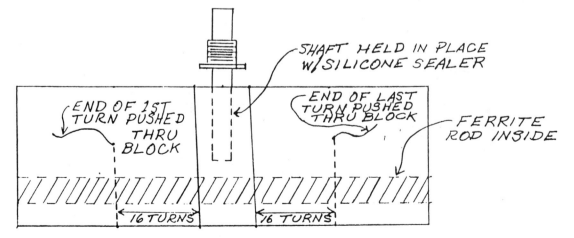

Fig. 5 -- Secondary coil.

using the crossed-loop/goniometer setup. These tests were made in my living room, so there were a lot of conducting objects in the vicinity of the loops.

Some further investigations are in order, obviously. I'd like to find out if preamps work as well at the goniometer position compared with how they work at the loops. Another item is that of providing an electrostatic shield between primary coils and the secondary coil. This is mentioned in Terman's description of the goniometer. An 80-meter version about 0.75 meter on a side would be worth looking into. I'm hoping to find time to investigate these and and some other ideas. Meanwhile, here's hoping to BCNU on 160!

References

\1 DeMaw, Doug, W1FB, "A Receiving Loop for 160 Meters"; **QST**, March 1974. (See also **The Antenna Anthology,** 1978, Chapter 4; and **The Antenna Book,** 14th edition.)

\2 Terman, **Radio Engineers' Handbook;** New York, McGraw-Hill, 1943; pp. 878-879.

\3 Henney, **The Radio Engineering Handbook,** 3rd Edition; New York, McGraw-Hill, 1941; p. 581.

\4 de Rosa, **Electronic Countermeasures,** Chapter 10; Los Altos, CA, Peninsula Publishing, 1978; pp. 10-2 to 10-5, 10-30 to 10-31 and 10-80 to 10-81.

\5 Cornell, **The Low and Medium Frequency Radio Scrapbook,** 3rd Edition; published by the Ham Radio Publishing Group, 1977; pp. 22, 44, and 48.

\6 DeMaw, "Receiver Preamps and How to Use Them"; **QST,** April 1984. (See also **The Antenna Anthology** and **The Antenna Book** 14th Edition.)

Bill of Materials

Item	Quantity
PVC pipe, 1/2 in x 10 ft, lawn irrigation type, **not** Schedule 40	4 pcs
PVC cross fittings, 1/2 in	3 ea
PVC T fittings, 1/2 in	4 ea
PVC 45-degree elbows, 1/2 in	16 ea
No. 4 self-tapping screws	44 ea (min)
RG-174 coaxial cable	80 ft
Aluminum chassis, 12 x 17 x 3 in, w/bottom plate (see text)	1 ea
Plastic refrigerator box (see text)	1 ea
Ferrite rod (see text)	1 ea
Styrofoam block, 12 x 12 x 3 in	1 pc
Plastic champagne cork	1 ea
Plastic milk jug tops	4 ea
No. 26 enamel-insulated wire	75 ft
Flexible twisted pair (see text)	2 ft
Very fine shielding braid (to cover above; see text)	2 ft
Small standoff insulators (type w/no. 4 mounting studs)	12 ea
5-lb. test nylon monofilament line	50 ft
Shielded/twisted pair	See text
Split-shot sinkers, no. 7	2 ea
Silicone sealer	1 tube
Silicone spray lubricant	1 can
Shaft bushing, 1/4-in dia. shaft	1 ea

Subsurface Antennas and the Amateur

By Richard Silberstein,* WØYBF

*3915 Pleasant Ridge Rd., Boulder, CO 80301

The term "subsurface" applies to buried or submerged antennas. Because of its general definition, it could also apply to a microwave antenna hidden in a meatloaf. There are two reasons for placing an antenna under ground or in the water: (1) concealment, and (2) hardening (protection against a nuclear blast).

Why would anyone want to conceal an antenna? An Amateur Radio operator with available attic space would be reluctant to uproot his yard. Subsurface antennas are used mainly for military and Civil Defense purposes, but amateurs should find them an exciting topic for experimentation and public service.

Scientists were thinking about buried antennas as early as 1912, and a loop antenna for submarines was proposed around 1919. Two instances of Amateur Radio experiments using this method were reported during 1948 to 1952. In the mid-50s, interest in buried antennas flourished, and many published papers appeared in the 1960s. After this period, interest died and the subject of subsurface antennas was put aside.

R. W. P. King and G. S. Smith recently published a book on subsurface antennas.\1 However, each paper I researched, each page I thumbed through, had one thing in common: They expound on antenna theory, but fail to give construction details. Why? Because nearly all the building, performance testing, and evaluation is done by contractors. This includes universities and "in-house" government groups. Much of this work goes astray, never reaching the journals. Either the results become proprietary (belong to the company that performed the research because of commercial potential), or they remain classified for security reasons.

Although theoretical literature on subsurface antennas is incomprehensible to most people, there is no reason to despair. The simplest antenna could never have been built if it had been necessary to understand the exact theory. Luckily, there is always some communication between the builders and researchers.

Thinking About Subsurface Attenuation

To help us relate to the theory of subsurface antennas, we must understand that much of what happens takes place near a boundary between two surfaces. Also, waves in a lossy medium weaken rapidly. Near-field wavefronts are curved, and so is the earth's surface. This causes the wavefront to tilt forward as it travels along the surface, radiating energy into the ground or the sea. To illustrate the result of applied theory used later in this article, calculations are made using the concept of a plane wave rapidly losing energy as it goes through ground or sea water.

To express attenuation loss, the decibel (dB) is used, where:

$$1 \text{ dB of voltage change} = 20 \log_{10} E_1/E_2$$

$$1 \text{ dB of power change} = 10 \log_{10} P_1/P_2$$

Some receivers are calibrated with S units of 5 dB each up to S9.

The term "skin depth" is often used when working with buried antennas. The skin depth of any soil is the depth at which the electric field is attenuated 1 neper (8.68 dB).

Skywave to Ground

Amateurs who work in the HF band are most familiar with skywave signals reflected from the ionosphere. At each receiving site, the pattern of the elevated antenna shown in Fig. 1 at P results partly from the combination of an incident wave A and a wave B' reflected from the ground. There is also wave C, the refracted wave, going into the ground. This is the one of interest for buried antennas such as the one at P'. In this case, the reflected wave is wasted. Note that the refracted wave is heavily attenuated.

To transmit from a buried antenna, start at P' in Fig. 1. The signal becomes skywave and travels toward the ionosphere along B so there is no energy loss to the reflected wave B'. At some angle, however, total internal reflection begins to prevail.

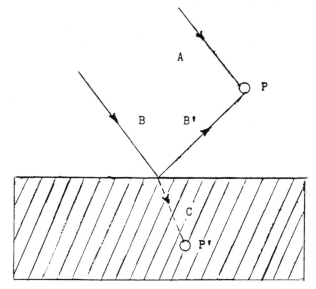

Fig. 1 — Reception above and below ground via skywave.

Incident- and Ground-Reflected Waves

Fig. 2 shows reception for short distances from any antenna. From a transmitter, T, there is a direct wave A and reflected wave B' combining at the receiver, R. For underground reception, the heavily attenuated wave, C, is still received at R'.

The behavior for Fig. 2 is qualitatively valid at all frequencies (except for some high-angle sky-waves at lower frequencies), but at distances not exceeding the line of sight. The received wave is called the "ground wave." The ground (also water), has a dielectric constant and conductivity that, acting together, cause its reflective properties to change with frequency. Its attenuation is also a function of frequency of the wave, and particularly of the distance traveled in the medium.

The Surface Wave

As distance increases, curvature of the earth becomes important. This curvature and the wave front give rise to a "surface wave" that becomes the sole groundwave beyond line of sight (see Fig. 3).\2,3,4,5

The surface wave is much stronger for vertical electric polarization (horizontal magnetic polarization) and attenuates less with distance at lower frequencies. At standard broadcast frequencies over average soil, a station may have a service range of 50 miles, with more for "good" soil. Low and very-low frequencies propagated as surface waves over sea water may be detected beyond a thousand miles.

Sea water has a dielectric constant of 81 and possesses high conductivity. It is vastly better than soil for surface propagation, but correspondingly worse for subsurface propagation at any depth. On the lower frequencies at great distances, it is the ionospherically reflected wave penetrating sea water that is the important mode for undersea reception.

As noted earlier, the surface wave in Fig. 3 tilts forward as it moves, radiating part of its energy into the ground or water. This same energy reaches a subsurface antenna. In transmission from below the surface, energy goes upward and contributes to the tilted wave. Beyond the surface-wave range, which may be relatively short distances in the HF band, energy penetrating the soil or sea must come first via skywaves.

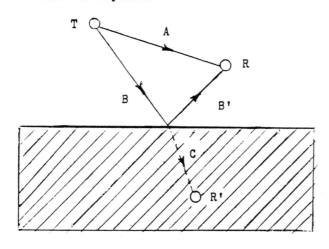

Fig. 2 -- Reception above and below ground from a source within line of sight.

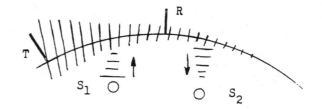

Fig. 3 -- Surface wave. T and R are a surface transmitter and receiver. S_1 and S_2 are submarines transmitting and receiving via the surface wave.

Estimates of Subsurface Attenuation of the Refracted Wave

It is difficult to estimate the attenuation of a surface wave in terms of depth below the surface, but easier to see what happens to the refracted wave from a downcoming skywave (Fig. 1). When working with skywave, important sources of field-strength loss are reflection at the point of entry, attenuation in going through the medium (in the refracted wave), and direct ohmic loss caused by currents going from the antenna structure through the earth or water.

Ohmic loss is also evident in broadcast-antenna ground screens. Insulation of ground-plane wires (if the ground plane lies on the ground or roof) was recognized by manufacturers of amateur verticals before 1960. H. A. Wheeler describes these losses in a more sophisticated manner and his papers should be studied for a better understanding.\6

Using the "propagation constant" of Appendix I, specifically the attenuation constant, α, for a simple plane wave, calculations were made to show qualitatively how attenuation loss varies with frequency and distance under different subsurface conditions. In plane-wave theory, it is assumed that a traveling wave in the earth or water will be attenuated exponentially. This means that starting at any point, the proportion of energy lost per unit distance is always constant.

Results show that in penetrating average soil at 5 MHz, a wave will be attenuated about 1.87 dB per meter of travel. The reflection loss must be added to this. With an average loss of 4 dB in the reflected wave, 40% is reflected and 60% must be refracted, or it is said that the refracted wave is down by 2 dB. Think about it.

In sea water the dB attenuation per meter is proportional to the square root of the frequency, at frequencies up into the lower VHF region. Lower frequencies, then, should show less loss in going through salt water.

At 20 kHz, the calculated subsurface propagation loss is only 5.4 dB per meter. However, for an arriving skywave, most of the power has already been lost upon reflection. Since salt water is a good reflector, perhaps there is a loss of 50 dB to the refracted wave. Now the total signal loss at 1 meter in depth is 5.4 dB + 50 dB = 55.4 dB. At a depth of 10 meters, it is 54 dB + 50 dB = 104 dB. At a frequency 100 times as great or 2 MHz, the attenuation would be 10 times as great or 54 dB at one meter. It would have a slightly smaller reflection loss, giving a total loss of about 100 dB at 1 meter in depth. This does not include antenna circuit losses. Two MHz doesn't look good for undersea communication.

In any subsurface reception, atmospheric noise and interference will be attenuated almost as much as the signal. The receiver noise figure limits reception, and a sensitive one can be used advantageously.

Spectrum space is limited at low frequencies. With weak signals, narrow bandwidths must be used for reasons of signal-to-noise ratio and spectrum space available. This reduces the speed of communications.

Some Practical Antennas

Practical Antennas for Very Low Frequencies

The VLF frequency range (3-30 kHz) and the LF range (30-300 kHz) could support reception of European broadcast stations, submarine communications, reception of standard-frequency and navigation-aid signals. Loop antennas, possibly in radomes, are logical antennas for submarines. On land, large conventional antennas or buried ones similar to those described under Medium Frequenices would be used.

As we have seen, direct propagation through salt water is highly attenuated. The predominant source of signal energy is a surface wave up to about 1,000 miles, with energy propagating to or from submarines to the surface-wave region. At long distances, skywave becomes the superior mode above the surface, with downcoming waves penetrating the ocean. The skywave is guided between the earth and the lower ionosphere.

At the time of this writing, a project is undergoing construction in Michigan using VLF. It consists of burying miles of underground wire in a forest area and will enable submarine communication to naval bases.

Medium Frequencies, MF (300-3000 kHz)

These are short-distance surface-wave communication frequencies, free of ionospheric disturbances. They are also used in transmitting and receiving information in the broadcast bands to and from emergency shelters. Some skywave communications are possible at night in the 160-meter band during times of minimum atmospheric noise. Ionospheric absorption on these frequency ranges are too great in the daytime.

Antennas for the MF are horizontal, insulated, buried wires fed in the center or at one end. Fig. 4 shows such wires that are grounded or insulated at the ends. The end-grounded wire has a high-loss return path to the feed point and resembles a cable with a lossy sheath. When the wire is near the surface, the largest return currents are below the surface so that the wire acts like an elongated vertical loop. It generates or receives a maximum surface wave off the ends of the antenna, and provides vertically propagated energy for feeding the forward tilt component of the surface wave.

The lossiness produces a low-Q antenna, but with resonance points visible as frequency is varied. This makes the antenna broadband, but inefficient. The ends can also be free, however, earth losses will be fairly large. Low-loss loop antennas are available that tune to the upper part of the MF band.\7

Wheeler suggests a more efficient buried antenna where the the straight long wire is replaced by a narrow insulated loop of the same length (Fig.

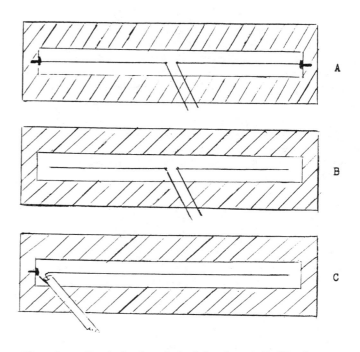

Fig. 4 -- Buried, insulated horizontal dipoles. At A, the dipole is center fed with the ends grounded. B displays a center-fed dipole with the ends free, and C shows one end fed, with the other free.

5).\8 This eliminates the return-current ground loss and would give the antenna a higher Q, allowing it to work best near resonant frequencies.

Refs. 6 and 9 give total losses for MF test antennas that differ markedly. Wheeler shows -28.5 dB for an antenna 1000 ft long and 3 ft deep. Fenwick and Weeks show as much as 10 dB more for an antenna nearer the surface, but only 1/10 as long.\9 The tests were performed on soil with different losses.

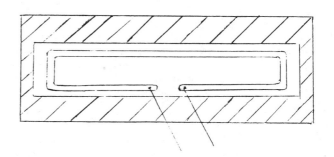

Fig. 5 -- An insulated, elongated loop as a buried antenna.

High Frequencies, HF (3–30 MHz)

Since skywaves are associated with high frequencies, buried antennas here would generally be used for fairly long-distance communications. Buried single dipoles, dipole arrays, loops and various other antenna configurations can be used in this band.

In 1965, I attempted to test the feasibility of using buried wires for communication when the objective was to conceal a tactical position. There was no doubt that this could be done; the question was whether something simpler would work as well or better.

The experiment was devised for reception of high-angle skywave from standard-frequency station WWV, then at Beltsville, MD on 5 MHz, at a distance of about 200 miles. A half-wave center-fed dipole (a doublet), insulated and free at the ends, was chosen. It was known that the effect of the ground is to reduce the physical length of the wire for resonance, because the velocity of propagation is reduced. To get some idea of the maximum required length for 5 MHz, the resonant frequency of a measured length of wire lying on the ground was noted. Now the length of a half wavelength of a wave in free space at 5 MHz is 492/5.0 = 98.4 ft. Because the velocity of propagation along a high thin wire is 95% of the free-space velocity C, a half wavelength of this wire at 5 MHz is determined by the well-known 468/5.0 procedure, yielding 93.6 ft.

For a wire lying on our particular ground, using a grid dipper and measuring resonance, the velocity was found to be 55.3% of C. Taking 55.3% of the above free-space half wavelength for 5.0 MHz, 98.4 ft, yielded 54.4 ft for the wire on the surface. When the antenna was finally insulated and buried in a trench 8 inches below the ground, the total antenna length for a 5-MHz half-wave resonance was 46.6 ft for a velocity factor of 47.4%. Some of the references show velocity factors as low as 25% for deeper antennas.

Both the buried antenna and a comparison doublet were placed across the path to Beltsville. Fig. 6 is a diagram of the buried antenna. Since the experiment was done before the time of PVC tubing and joints, the wire was put into vinyl hose for insulation (not an original idea); the wire and the hose were cut into two equal lengths with the wire shorter than the hose. Each length of hose was cemented at one end around a hole in a vinyl trash container with a lid, which was located in a sump at the middle of the trench. Inside the container there was a smaller metal can containing the balanced matching network. Feedthrough insulators were provided at each side of the metal can for connection to the antenna wires as well as a standard socket for the 50-ohm output cable, near the top. Nylon strings were attached to the ends of the no. 12 soft-drawn copper wires and brought up to the surface through the bent-up ends of the hose. Because we started with longer wires than necessary, this allowed us to shorten them to the proper length, keeping them taut.

Precautions were taken to see that all joints were waterproof and that there was some room in the sump for water accumulation. A cable (RG-8) was brought out the top of the sump and the area nearby was covered with a piece of galvanized sheet metal.

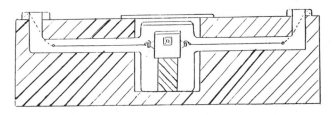

Fig. 6 — The half-wave dipole and matching unit in 5-MHz tests. The first version used vinyl hose for insulation. Better results should be available using plastic tubes, spacers and water-tight fittings.

Comparison of reception was made using a half-wavelength horizontal doublet 0.3 wavelength above ground. Both antennas were positioned across the path to Beltsville, and not too close to each other. The balanced and matched cables from each antenna went to a test shelter. A stable high-quality communications receiver was calibrated in decibels relative to one microvolt, using a good-quality signal generator and a strip-chart recorder fed by the AGC voltage. The recorder was an old Esterline-Angus galvanometric type. Sometimes a bridge-balancing or digital recorder should not be used with a narrow-band receiver, especially if the receiver and signal generator are unstable and require manual tuning. Some recorders will not respond quickly enough to show when the receiver is resonated.

The test consisted of switching antennas every 5 minutes for two hours with the recorder running. The most constant difference was about 16 dB of loss in the buried antenna, both in the signal and in the noise during station-off periods. To ascertain that the signal was not being contaminated by pickup through the cable shielding, the cable was unplugged at the buried antenna and terminated in 50 ohms: The signal strength fell 41 dB.

With these results, it was concluded that communications with a site using a buried antenna was certainly possible. If this method is used for concealment, something simpler and easier to install might be desirable such as a loop or a disguised loaded whip.

I recently noted the existence of a report describing experimental work done several years later on buried antennas at the Rome Air Development Center (RADC).\10 The report gives theoretical background and describes some careful, detailed research on buried antennas.

The buried wires in the RADC experiment were laid in trenches much as described for the 5-MHz tests above. The insulated wires were actually RG-19 cable without the outside shield. One difference was that several wires were buried so that it was possible to obtain some small gain by paralleling pairs separated by enough distance to make mutual coupling insignificant. Separation of 13.8 ft gave a distance of 1.4 "skin depths" near 7 MHz for their soil constants. One advantage of the RADC construction method was that solid insulation and an above-ground matching network eliminated water-accumulation problems, although insulation by wide-diameter air-filled tubes might have reduced ground losses.

Pulse and CW signals with a dipole pair buried 3 ft afforded surface-wave signals at 7 MHz out to 40 miles under favorable conditions of low noise.

Power was probably only 22 W. The signal strength was 18.2 dB worse than for a 35-ft whip on the surface. Similarly, under favorable absorption and noise conditions, skywaves were seen out to 500 miles and were shown to be 24 dB weaker than for a half-wave dipole 1/4-wavelength above ground.

Results and theory were combined at RADC to determine the performance of an array of buried dipoles, one half free-space wavelength long and 3 ft deep. With a transmitter power of 10 kW, received signal conditions were as follows:

1. For surface waves at 2 and 5 MHz under bad atmospheric noise conditions at night, the signal was down to about 10 dB above noise at around 100 mi. Bandwidth was not stated.

2. A similar antenna at 7 MHz for skywave under winter daylight conditions at a sunspot number of 100 had a signal equal to about 10 dB above noise at 1000 mi. Bandwidth was not stated.

3. Winter-night skywave at 2 MHz had a range over 2500 mi for 10 dB signal to noise. Bandwidth was not stated.

HF Dipole Antenna Characteristics

Plots of antenna resistance and reactance in Refs. 9 and 10 show resonant frequencies which indicate a velocity of propagation of only 25 to 30% of C.

In Ref. 10, consider E_ϕ , the horizontal broadside electric field (TE mode), and E_θ , the off-end "vertical" electric field (TM mode) pattern. For a buried dipole at a given frequency, these patterns are similar in shape to those for a half-wave doublet a quarter wavelength above ground. With a buried dipole, the TM-mode pattern shows a stronger response than does the TE-mode pattern as the angle above the horizon decreases.

The current along a lengthy, buried dipole, may exhibit resonances. Because of losses, however, it shows much less of a standing-wave pattern than does an above-ground antenna. Also, because of the slow velocity of propagation in the medium, these standing waves are crowded close together. Therefore, not much of a lobe structure above ground is created. This apparently means that for a velocity factor on a wire in soil of 25%, a given dipole should be able to operate on frequencies up to values corresponding to four half wavelengths in soil without a bad antenna pattern. Of course, the antenna must be properly matched to accept power. In Fig. 3-46 of Ref. 10, there are some dips in the field strength-versus-frequency curve at $3\lambda/2$ and 3λ which are not easily explained. There is obviously never any vertical lobe structure caused by height above ground!

Other HF Antennas

An intriguing antenna for concealment and partial hardening is a high-Q tuned loop. Those made by one manufacturer have about an 8-to-1 tuning ratio and can be used for transmitting.[11] A recent compact model, the MLH-2/D is a half-loop on a ground plane. The manufacturer feels that these loops would not function well if buried in the ground, but could be used in a metal-lined conical cavity with protective dielectric lid or, less expensively, in a wire-mesh lined crater covered with such a lid. Considering a claim made many years ago for an early version of this antenna, only 3 dB worse than an average outdoor doublet at 4 MHz, I wonder if it might not perform well in a large subterranean cavity.

For obtaining gain, the usual Yagi or log-periodic antenna would have to be installed in an enormous subsurface room. However, it is much simpler to obtain gain in buried antennas by reducing loss. This is done by paralleling dipoles as noted above. Another feasible way to obtain gain would be to design a buried V or rhombic antenna. Perhaps inserting capacitors in the wire periodically could increase the phase velocity to simulate a wire above ground. Wheeler has mentioned this technique for acquiring a uniform current in buried wires; others have suggested it for above ground antennas.[12]

It is suggested that experimenters try dipoles with more insulation than has been used. Wide-diameter PVC water pipes and even plastic sewer pipes suggest themselves. Internal supports would be needed and waterproofing is essential.

VHF Antennas

VHF antennas can be employed just below ground, the main purpose being concealment. A study made near 2 meters and above was concerned with communication at low angles to aircraft and nearby fixed stations.[13] As expected, the TM mode was the most efficient in generating this surface wave.

A full, insulated horizontal dipole and a variety of miniature tuned loops and dipoles mounted a few centimeters deep produced TM fields at angles of about 10 degrees above the surface that were 14 or 15 dB weaker than TM fields from a test quarter-wave monopole on the ground. Considering cosmic noise as the only competition, it was estimated that with such an antenna and only one watt of power, communication with aircraft was possible at 50,000 ft to distances of 70 to 90 mi.

An antenna not covered in the references is the slot antenna. It is not very difficult to construct at VHF and might be mounted just below a plastic lid on the ground. Beneath the antenna a mesh-lined cavity could exist. Arrays of resonant slots could offer substantial gain. It is worth noting that reception has been possible with a 2-meter hand-held radio in a plastic bag under salt water to a depth of 20 ft.[14]

Buried Antennas in a Nuclear Emergency

Although the peril of a nuclear attack is "unthinkable," most people would want to survive for as long as possible. The topic of survival communication then is something many will want to consider.

Because of the great attenuation of radio waves associated with a deeply buried antenna, Civil Defense (FEMA) personnel apparently now lean toward pop-up monopole antennas capable of pushing aside a ton of debris. These would be used only after an attack had ceased. This policy raises the question of how survivors would learn, in a deep shelter, that the attack had ceased. Some of the loop antennas described above are installed at Civil Defense locations in the recommended manner that would be of assistance in this situation.

One problem that has not been mentioned is that of EMP; the short, powerful electromagnetic pulse from a nuclear blast.[15] Under certain high-altitude conditions, explosions can disable a receiver a thousand miles away. It would be wishful thinking to hope there is no solution to EMP. Unfortunately, such a solution must exist, otherwise one could relax in the comforting thought that if missile-guidance systems would fail because of EMP, there could be no sustained nuclear war. Maybe some day everybody will be able to buy a kit which will "EMP-proof" computers, televisions, and radio equipment -- a poor compromise, but possibly realistic.

Consider that only a small area of a country is in the direct-hit category. It becomes obvious that the great difficulty of total hardening of a communications antenna without destroying its effectiveness should not deter effort to install antennas which, though only partially hardened, would provide some useful communications in the vast majority of cases.

The Role of the Amateur

Many years ago, I noted that it was possible to hear 20-meter daytime DX signals with a two-tube regenerative receiver connected to a short wire in the basement. Signals were received, even if they might have been reradiated from power lines, pipes, and heat ducts. For most people in a nuclear emergency, the basement would be the logical shelter. A few others would have bomb shelters. In either case, an amateur should be able to communicate with, say 10 watts of battery power, using narrow receiver bandwidths and hand-keyed CW. The antenna could be a very high-quality loop, a wire strung in the basement, or a center-fed long wire, insulated, ends not grounded, and buried just under the ground. An adjustable matching network might be desirable, especially for a wire buried in the soil, because impedance changes with frequency as well as with soil moisture.

Amateurs could possibly assist Civil Defense units by conducting experiments and tests. The idea of a loop in a large underground room is intriguing; one would have to consider the possible need in some cases of filtering out strong broadcast- or power-frequency fields causing front-end overload and intermodulation. Common-mode pickup in certain types of balanced loops could be a problem.

APPENDIX I

Attenuation of Plane Waves in Subsurface Media

An understanding of the material presented here is not necessary for a general application of the problems of subsurface radio communication. However, for those whose science education has taken them through Maxwell's Equations, use of a plane-wave analytical approach should make it possible to have a more quantitative understanding of the effect of soil or water on wave attenuation as distance, frequency and constants of the media are varied.

For **exact** methods of solving the problems of subsurface antennas and propagation, the equivalent of a graduate science background and knowledge of complex electromagnetic problems is required.

Many textbooks show that in the propagation of a plane electromagnetic wave in a homogeneous medium along the x axis, Maxwell's Equations yield the following for the electric field:

$$E = E_o e^{\pm \Gamma x} \qquad \text{(Eq. 1)}$$

where, (in the MKS system of units)

x is the distance in meters

E_o is the electric field in volts per meter at x = 0
E is the electric field at any distance along the x axis
e is the Napierian base
Γ is the propagation constant

For attenuation and phase retardation, the negative sign in (1) is taken so that:

$$E = E_o e^{-\Gamma x} \qquad \text{(Eq. 2)}$$

$$\Gamma = j \omega \sqrt{\mu \varepsilon (1 + \frac{\sigma}{j \omega \varepsilon})} \qquad \text{(Eq. 3)}$$

where,

$\omega = 2 \pi f$
$\varepsilon = K \varepsilon_o$

f is frequency in Hz
k is relative dielectric constant
$\mu = \mu_o$ for non-magnetic material

$\varepsilon_o = 8.855 \times 10^{-12}$

$\mu_o = 1.2566 \times 10^{-6}$

σ is soil conductivity, mhos/meter

Since Γ is complex, it can be written:

$$\Gamma = \alpha + j \beta$$

so Eq. 2 becomes:

$$E = E_o e^{-(\alpha X + j \beta X)}$$
$$= E_o e^{-\alpha X} \times e^{-j \beta X} \qquad \text{(Eq. 4)}$$

α is now attenuation in nepers per meter. (1 neper = 8.68 dB) and β represents phase retardation per meter. In free space α is zero and $\beta = 2 \pi / \lambda$. However, we will be interested only in α.

There is a problem in reducing (3) to obtain the value of α because one runs into a 4th-degree equation. A solution appears in Ref. 16, p. 271, Eq. 3. However, using De Moivre's Equation yields the identical solution. [17] Adapting that equation to the square root of a complex quantity:

$$\sqrt{A + jB} = \sqrt{\rho} \, (\cos \theta /2 + j \sin \theta /2) \qquad \text{(Eq. 5)}$$

where

$$\rho = \sqrt{A^2 + B^2}, \quad \cos \theta = A/\rho \quad \sin \theta = B/\rho$$

For convenience we now write (Eq. 3) as:

$$\Gamma = j \omega \sqrt{\mu \varepsilon} \sqrt{1 - \frac{j \sigma}{\omega \varepsilon}} \qquad \text{(Eq. 6)}$$

and substitute the term under the second radical in Eq. 5.

For average soil, the constant is σ = 0.005 mhos/meter, k = 15.

For sea water, σ = 5 mhos/meter, k = 81.

References

\1 King, R. W. P. and G. S. Smith, **Antennas in Matter: Fundamentals, Theory, and Applications,** MIT Press, 1981.

\2 Burrows, C. R., "The Surface Wave in Radio Propagation over Plane Earth," **Proceedings IRE 25,** p. 219, 1937.

\3 Jordan, E. C. and K. G. Balmain, **Electromagnetic Waves and Radiating Systems,** 2nd edition, Prentice-Hall, 1968.

\4 Norton, K. A., "The Physical Reality of Space and Surface Waves in Radiation Fields of Radio Antennas," **Proceedings IRE 25,** p. 1192, 1937.

\5 Norton, K. A., "The Calculation of Ground Wave Field Intensities Over a Finitely Conducting Spherical Earth," **Proceedings IRE 29,** p. 623, 1941.

\6 Wheeler, H. A., "Useful Radiation from an Underground Antenna," **Journal of Research of the National Bureau of Standards,** Vol. 65D, p. 89-91, January-February 1961.

\7 Miniloop HF Tunable Antenna Model MLA-1/E 1.8-14.5 MHz for transmitting and receiving. For bulletins on this and other loops, write: Antenna Research Associates, P. O. Box 196, Beltsville, MD 20705. (One of ARA's loops is described in **CQ Magazine** for Jan. 1973, p. 84, White's column, "Surplus Sidelights.")

\8 Same as no. 6.

\9 Fenwick, R. C. and W. L. Weeks, "Submerged Antenna Characteristics," **IEEE Trans. AP-11,** p. 296, May 1963.

\10 Entzminger, J. M. Jr., T. F. Treadway and S. H. Talbot, "Measured Performance of Subsurface Dipoles," RADC-TR-69-221, Technical Report June 1969. This report should be available for a fee from the National Technical Information Services (NTIS) under no. AD690609.

\11 See Ref. 7.

\12 See Ref. 6.

\13 Bussey, H. E. and E. B. Larsen, "Buried Antenna Performance. Development of Small Resonant Buried Antennas," (work done at National Bureau of Standards, Boulder, CO.) Report RADC-TR-74-169 Final Report June 1974.

\14 West, G., WB6NOA, "Maritime Radio," **World Radio Magazine,** March 1984, p. 30.

\15 Three articles on EMP appear in ITEM-84 (Interference Technology Engineers' Master), pp. 36-50. This is a trade journal published by R & B, Enterprise Division of Rober Industries, 20 Clipper Road, Consohocken, PA 19428.

\16 Ramo, S. and J. R. Whinnery, **Fields and Waves in Modern Radio,** Wiley (New York) and Chapman & Hall, (London), 1944.

\17 **Reference Data for Radio Engineers,** 5th edition, Howard W. Sams & Co., 1968, p. 44-4.

Antenna Construction
and Installation

A New Approach to the Construction of Large Yagi Beams

By Arie Bles,* VK2AVA

*78 Chapman Parade, Faulconbridge, Australia 2776

Anyone who has ever contemplated on constructing a Yagi beam for frequencies lower than 14 MHz has experienced problems in obtaining building materials. The required boom length for a 3-element beam is at least 30 ft. A strong, large tower and rotator are musts, and the material used for a full- or near full-size element is physically heavy.

There are various methods to reduce the element lengths. Hy-Gain and KLM do it with "linear loading." Linear loading means that the center part of the element is folded back on itself. This method creates a sacrifice in gain and bandwidth because current is greatest in the center of the half-wave element, and most radiation occurs at this point. A better method can be to place a loading coil on both sides of the center towards the ends. Because this will weight the elements, heavier and thicker material for element construction is necessary. A shorter boom length cannot be used if you are to retain bandwidth and gain.

During a visit to Germany some years ago, I observed a demonstration using a fiberglass fishing rod. The demonstrator had mounted a small 2-meter Yagi near the thin end. To my surprise, he lifted the rod in a dangerous manner from a horizontal to vertical position without incident!

The fishing rod had eight 3-ft 6-in sections, telescoping to 25 ft total length, with a total weight of under 2 lbs. It was constructed of flexible, thin-walled material about 1.5-in diameter, and it was possible to bend this rod almost 90 degrees without damage. The rod returns to its original straight condition as required for the sport of fishing.

I had a vision of using this fiberglass rod as a full-size 40-meter element with a thin wire fed through it. It could be connected to 20 ft of 1-3/4-inch aluminum tubing, giving an overall length of 70 ft and resonating on 40 meters!

These 25-ft fishing rods with 1/8-in aluminum wire through them are now the basic parts of my 3-element full-size Yagi for 10 MHz. With a 20-ft x 2-in boom and brackets to mount the elements, it gives me a full-size 10-MHz Yagi with the total weight of a 21-MHz beam!

Element center-mounting U-frame, showing insulation blocks, 4-in muffler clamps and the center of the 1-3/4-in tubing, and collapsed 8-section fiberglass fishing rods.

For 40 meters, I use six of the same kind of fishing rods, but they are placed into the ends of 20 ft of aluminum tubing. The 1/8-in wires are connected to the tubing ends through the fishing rods. Each element has to be carefully resonated, as they are a bit longer than full aluminum tubing elements. They are similar to tapering elements, thick in the center to thin at the ends.

Other problems were gathering the boom material and brackets for the elements so I could mount them on a boom. Two-inch material for the boom is too weak. I am presently using thin-walled 4-inch irrigation tubing of only 1.2 mm wall thickness. It is very light, but strong enough for an 11- to 12-meter-long boom, provided one uses 4-inch full-circumference muffler clamps for element and boom-to-mast mountings. I don't recommend using a CDR HAM III or IV rotator. A heavier-duty one such as a prop-pitch motor is essential.

I use lengths of 4-ft aluminum U-frame material for element-to-boom mounting brackets. They are 4-in wide and 1-1/2-in deep. Four blocks of insulating material were needed to mount the element tubes on the U-frame; the two in the center are somewhat smaller than the other two at the ends. (I used wood

boiled in paraffin. This method was used 30 years ago for twin-feeder spreaders. Any non-hygroscopic material can be used.) It gives a gull-wing appearance, but counteracts element sagging. The center ones are spaced sufficiently apart to accommodate the muffler clamps for attachment to the boom.

It is handy to break the elements in the center to establish correct resonance with a dip meter or similar method. A heavy strap in the center of the parasitic elements provide for continuity. The driven elements are fed with 50-ohm coax cable and a 1:1 balun. I accept the mismatch and standing wave on the feed line, but there are methods such as the Hy-Gain beta match to reduce that.

All materials except the fishing rods were obtained locally. The rods were imported directly from Germany and cost around 30 U.S. dollars each. The total weight of my 40-meter full-size beam is less than 70 lbs, and many amateurs worldwide have been impressed with the signals it produces. Strong winds during the 18 months of my using the antenna have not caused any damage. I tested rod lengths for cold climates by deep-freezing them. They remained flexible when removed from that environment.

The 40-meter beam. The tower to the right is a stacked homemade 204BA over the 3-element 10-MHz Yagi.

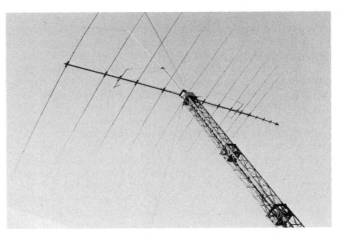

My 14-element log-periodic beam covering 14 to 30 MHz.

Raising Beam Antennas

By Lawson Young,* N4LY

559 E. Wesley Rd., Atlanta, GA 30305

The installation of large beam-antenna arrays on tall towers has always been a problem. I have found the pilot-catenary back-guy (PCBG) method to make this task simple. Whether the tower is guyed or free standing, against a structure or located in trees, raising a beam can be easy.

The only requirement for using the PCBG technique is that you have a large, clear space on the ground so the assembled beam antenna can lay in its normal mounted position. The area should be clear for the antenna to pass through from the ground to the top of the tower. This may sound complex, but the entire raising or lowering of the beam can be performed by one person. Of course, the job is easier with a crew of more than one.

The PCBG method is an old material-handling technique. Its most visible use is in the logging industry where heavy logs are transported through rough terrain. One cable is installed at a rather shallow angle and a trolley (pulley) is attached so that the material can be moved up and down. The second cable, the back guy, is a guy wire that reinforces the first cable from the back.

Before you decide to use this technique, make some preliminary measurements of the space you have to work with. If room is limited, you will have to raise your beam another way. Most 100 x 200-ft lots will accommodate this procedure. For smaller lots, consider using a portion of your neighbor's yard. The use of this space is temporary and the most you will do on their property is drive a stake in the ground which is removed after you complete the job. An idea of the cable arrangement to determine needed space is shown in Fig. 1.

The pilot catenary and back guy are exactly opposite each other, 180 degrees apart. The back-guy anchor point can be 40% to 50% or greater of the tower height from the base. The pilot-catenary anchor point from the tower base should be 100% to 150% of the tower height. One hundred percent of the height gives you a 45 degree angle. You should not try to obtain a larger incline than that. About 125% of the tower height is optimum.

Once you have established that there is room for the pilot catenary and back guy, be sure you have clearance for the antenna to move up the pilot catenary to the top of the tower. Careful visual checks will determine this.

After you are sure of the anchor points and clearance, installing the cables and driving the stakes completes the job. One additional concession: Be sure you have a mast pipe that extends two to six ft above the top of the tower. See Fig. 2. A standard 10-ft mast section mounted in your rotor at the top of the tower gives a 6- to 8-ft clearance above the top of the tower. If you do not want to use your mast for this purpose, mount a temporary gin pole securely to the top for attachment of the pilot catenary and back guy.

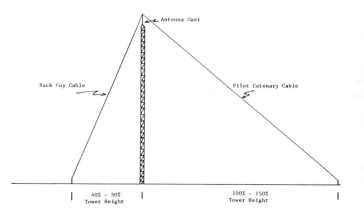

Fig. 1 — How to calculate anchor-point distances.

Two 1/4-in eyebolts or hooks can be installed at the top of the mast to secure the pilot-catenary and back guy. These are permanent and will be out of the way. If you use a temporary gin pole, install the hooks at the top. Using 3/16- or 1/4-in stranded steel cable, attach the cables from the hooks at the top of the mast (or temporary gin pole) to the anchor points previously selected. The anchor points can be a tree, secure post or a 1-1/2-in pipe or rod driven 3 to 4 ft in the ground. The cables should be attached and drawn as tight as possible by hand. Cable clamps should be used to secure the cable. If a hoist or "come-along" is used to tighten the cable, be sure it is not too tight as the mast might be bent once the pilot catenary and back guys are in place. Temporarily drop the top set of tower guy wires, if in place, as the PCBG holds the tower very steady. Usually for a large beam, the set of guy wires at the top of the tower will be in the way. Disconnect them; dropping one guy wire will probably suffice. At most, two of the guys will have to be dropped.

A pulley that will fit the pilot-catenary cable should be installed before the cable is tightened to its anchor point. The pulley should have an eyelet or hook and it will be upside down because the antenna assembly hangs from it. This can be seen in Fig. 3. A rope, 3/8 or 1/2 in, should be tied to this pulley eyelet, run to the top of the mast on the tower, through a second pulley and then to the base of the tower. For a hundred-foot tower, this will take approximately 250 to 275 ft of rope. This rope is used to pull the beam assembly to the top of the tower.

The next step is to get your beam ready to be pulled up. The beam should be assembled and tested as would normally be done prior to installing it on any tower. A short temporary mast is placed in the normal mast position of the antenna, with one exception. The mast can be any convenient length, 1 to 3 ft. If your antenna has boom guying, then the temporary mast should be long enough to allow the boom guys to be attached as they would ordinarily.

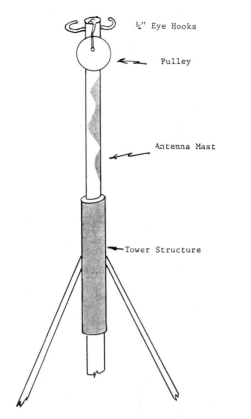

Fig. 2 -- Arrangement of antenna mast with hooks and pulley.

¼" Eye Hooks

Pulley

Antenna Mast

Tower Structure

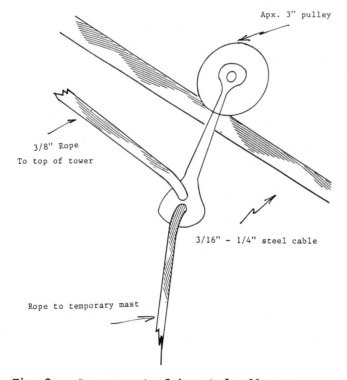

Apx. 3" pulley

3/8" Rope
To top of tower

3/16" - 1/4" steel cable

Rope to temporary mast

Fig. 3 -- Arrangement of inverted pulley or trolley on pilot-catenary cable with pull rope.

Provisions should be made at the top of this mast to secure a line or hook so that it can be connected to the pulley that is on the pilot catenary. A short piece of rope works fine, as this allows some flexibility in adjusting the distance from the pulley to the top of the temporary mast. This distance should be adjusted to match the distance from the top of the mast on the tower to the point above the tower top where the antenna will be attached. This is not critical, but should be considered.

To secure the temporary mast to the antenna boom-to-mast plate is simple, but slightly different. The mast is normally mounted to the boom-to-mast plate, on the side opposite from the boom. Naturally, they could not be mounted on the same side. The mast is usually held to this plate by one or two bolts above the boom and one or two below. To mount the temporary mast, put it on the same side as the boom, using only the top U bolts to fasten it. These top bolts are sufficient to hold the antenna while it is being hauled to the top of the tower. See Fig. 4. This leaves the opposite side of the boom-to-mast plate clear and ready to attach to the regular mast, using the bottom U bolts. The bolts will hold the antenna in place while the temporary mast is removed and the top U bolts installed on the regular mast.

When you have the temporary mast attached to the antenna, and connected to the pulley on the pilot catenary, you are ready to haul it to the top. You should be sure all the cables, clamps and connections are tight and secure. The antenna can float free as it is ascending, but guidelines can be connected to turn the beam if necessary. Be sure you have a way to disconnect these guidelines if used because the end of your antenna is difficult to reach from the ground. Care should be used in pulling. For safety, a second person should snub the rope around a post or tree in case the rope slips out of your hand. The antenna can be raised and lowered with one person, but for safety's sake, two or more people should be employed. A person on top of the tower while the antenna is being raised is recommended because he or she can watch for any obstructions the antenna might encounter. The antenna suffers no ill effects from this technique. It is subjected to less force than it will encounter during strong winter winds. See Fig. 5.

Once the antenna is intalled at the top, the lowered guy wires can be restored and the pilot catenary and back guy cables removed. The preparation for this technique takes a little time and thought, but the actual raising and installation of the antenna can be done in less than an hour.

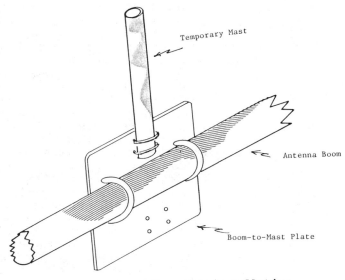

Temporary Mast

Antenna Boom

Boom-to-Mast Plate

Fig. 4 -- Temporary mast installation.

Fig. 5 -- Thirty-two-ft boom antenna, in
transit, with temporary mast in place and
suspended from pilot catenary.

General Antenna and Transmission-Line Information

The Horizontal Dipole Over Lossy Ground

By Robert B. Sandell,* W9RXC

*26 G. H. Baker Dr., Urbana, IL 61801

Technical material available to radio amateurs view antennas as wires hanging in free space, or above a ground of infinite conductivity. While these idealized conditions are necessary for the teaching of radio physics, they are of little help to the amateur trying to understand, or evaluate, his own real-world antenna. This paper, in a qualitative fashion, examines some of the effects of operating a horizontal dipole over a ground of finite conductivity. The structure of the near field and the far field are discussed briefly, and then four parameters of the horizontal dipole are examined as the antenna is moved from an environment of free space to one of close proximity with lossy ground. Driving-point impedance is first examined and a corrected curve is developed for radiation resistance versus height. Efficiency is explored next and graphical presentations exhibit heat losses incurred as the near field contacts the lossy media. In connection with efficiency, changes in bandwidth are noted. Finally, the vertical angle of radiation is calculated as a function of height and the reflection coefficient calculated as a function of soil constants.

Comments such as, "I put up a new antenna but it doesn't work like the **The ARRL Handbook** said it should," are frequently heard on the air. This is usually followed by a condemnation of antenna theory as worthless gibberish because it seems to deal with horizontal antennas in free space or vertical antennas above a perfectly conducting ground plane. What these well-meaning enthusiasts fail to realize is that antenna physics is divided into two parts. The first deals with antennas located in idealized environments (free space and above ground planes of infinite conductivity), while the second part deals with the effects that occur when the antenna is moved into a hostile environment (lossy ground). It is regrettable that the subject has been given such cursory treatment in amateur literature since real-world antennas can, at times, be profoundly effected [sic] by ground presence.

The following discussion deals, in largely qualitative fashion, with the effect of a lossy (real life) ground on four important parameters of horizontal antennas. Detailed calculations of ground effects are, as we shall see, largely impossible to make. An understanding of the basic ground proximity mechanisms, however, should enhance the ability to design antennas or to interpret their results.

Ground effects are not the same for vertical and horizontal polarization.\1 Although the remarks and data contained herein refer to single horizontal dipoles, they apply equally well to all horizontal antennas that are dipole derived. Uda-Yagi arrays, collinear beams, inverted V's and horizontally polarized cubical quads and delta loops fall into this category.

Earths and Fields

Soil or earth appears almost purely resistive at very long wavelengths. At very short wavelengths, it presents itself as an almost pure dielectric. In the so-called high-frequency range of 10 to 160 meters, earth takes on a combination of the two, or more simply, appears as a lossy dielectric.\2 This is unfortunate because it greatly increases the complexity of mathematical calculation. Earth can be categorized by parameters of permittivity and conductivity or the ability to store and conduct charge. Permittivity is generally specified in relative units, and conductivity in millisiemens per meter. Some typical values are:

	Conductivity	Permittivity
Good Soil	15	20
Average Soil	10	15
Poor Soil	2	10
Residential & Industrial	1	4

Maps containing regional conductivity figures are readily available; however, they are based on averages made from wide ranging data and thus are totally ineffective in pinpointing the conductivity at a specific location.\3,4 In addition, conductivity can vary both daily and seasonally. Maps give no hint as to dielectric qualities. Conductivity may be measured with some difficulty, but the measurement of soil dielectric qualities is beyond the scope of Amateur Radio.\5 It is for these reasons that most operators have only scant knowledge of their own soil parameters. The above table will serve as a fair estimate to use for evaluation purposes. In all probability, a majority of Amateur Radio sites will approach the residential-industrial category. While this may seem unduly stringent, it must be remembered that the original prairie contained no roads, sidewalks, or a host of other obstacles to good ground performance. Also, most urban and suburban areas usually have been subject to filling, scraping and leveling, all of which serve to alter ground characteristics.

Fig. 1 describes the fundamental geometry of a dipole operating over lossy ground. Point A represents the end of the antenna (the wire is perpendicular to the page). Region A is air, Region B is lossy ground, and the x axis is the air-earth interface. H is the height of the dipole above ground.

The area immediately surrounding an antenna contains not only the radiating field, but also strong reactive or storage fields. This stored energy is returned to the antenna at a later part of the cycle.\6 The strength of the radiating field falls off inversely to the distance from the antenna, while the stored energy falls off at rates inverse to the square and cube of the distance. At a distance of one-half wavelength, the stored energy has fallen to an insignificant value. At one-sixth of a wavelength, the radiation and storage fields are of equal strength.\7 In Fig. 1, r_1 has a length of one-sixth wavelength and r_2 has a length of one-half wavelength. The area within the circle whose radius is r_2 is known as the near field.

When a resonant half-wave dipole is operated in the environment of free space, its current and voltage will have a fixed ratio which is, by definition, an impedance. Since the current loop is coincident with the terminal points, it is also the radiation resistance. Traditionally, this ratio has been given as 73.13 ohms. (Later research indicates that at true resonance the figure should be 68 ohms, but the old value is still most commonly used.\14)

A change occurs when the dipole is moved into the earth's environment. A portion of the radiated energy is reflected by the ground and in turn is intercepted by the antenna. This intercepted signal causes a second current to flow in the antenna and thus modifies the original voltage/current ratio. This new ratio is known as the driving-point impedance. It is this impedance, which is effected by ground proximity, that we are concerned with in feeding and matching the antenna.\8,9 The solid curve in Fig. 2 is a commonly published representation of the driving point (radiation) resistance of a horizontal half-wave dipole at various heights above a **perfectly conducting** ground. It is obvious from examination of the curve that the greater the height, the less profound the ground effect on impedance. If the curve were extended to about ten wavelengths, it would be difficult to distinguish from a straight line.

Waves incident on a perfectly conducting surface are 100% reflected. This is not true for a lossy medium, such as earth, where a portion of the energy is absorbed. When the dipole is a long way from earth, the effect of its reflected signal is insignificant. As it moves closer to earth, the impedance variation becomes greater and greater. Eventually the near field will come into contact with the lossy earth. Stored energy is then absorbed by the earth and converted to heat instead of being returned to the antenna for later use. The dashed-portion curve in Fig. 2 represents the impedance variation when the near field comes into contact with a poor-quality ground.\10,11 This phenomenon is known as near-field coupling. As the quality of the soil improves, the curve approaches the solid-line value. This effect of height versus impedance is especially important when evaluating pre-cut or pre-tuned antennas or in tuning arrays at ground level.

Efficiency

The efficiency problem versus height above ground is closely related to the impedance problem; both involve the concept of near-field coupling. When the antenna is high above ground, the amount of power lost to ground heating is insignificant. As the near field, with its large quantity of stored energy, comes into contact with lossy earth, power is lost that would normally end up as radiated power.\12 Fig. 3 illustrates the estimated fall-off of efficiency as a dipole, operating at 18 MHz, is moved close to a poor-quality earth. Note that below

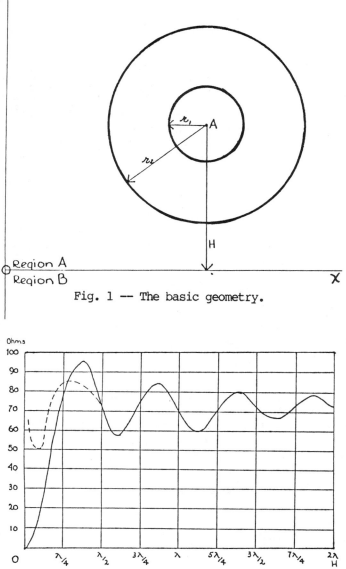

Fig. 1 -- The basic geometry.

Fig. 2 -- The driving-point resistance of a half-wave dipole over a ground of infinite conductivity and over lossy ground.

one-fifth wavelength, the curve falls off monotonically toward zero. While good electrical height is easily attained at high frequencies, it comes at a high cost below about 5 MHz. Those who contemplate antennas for the 80- and 160-meter bands should certainly consider the trade-off between efficiency and the cost of vertical support structures.

Bandwidth

Bandwidth is defined as the frequency range in which an antenna parameter will hold to within a given limit.\13 The parameter may be impedance, pattern or gain. The most frequently used parameter is impedance, as indicated by the standing wave ratio. Thus, the impedance bandwidth would be the band of frequencies where the antenna exhibits an SWR of less than, or equal to, some fixed limit (often 2:1). Bandwidth is an inverse function of Q (the figure of merit) which, in the case of an antenna, is the ratio of the ability to store power to the ability to radiate power.\14 Expressed in more definitive form:

$$\text{Bandwidth} \propto \frac{1}{Q} \propto \frac{\text{Power Radiated}}{\text{Power Stored}}$$

A portion of the storage field is dissipated when it interfaces with lossy earth. As stored power is reduced, Q is reduced, resulting in increased bandwidth. While wide bandwidth is a desirable characteristic, in this case it is accomplished at the high price of efficiency. Some misleading bandwidth figures are sometimes published because an experimental or test antenna was located too close to a lossy ground.

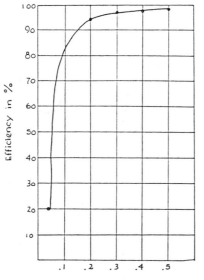

Fig. 3 -- Efficiency of a half-wave dipole over poor quality soil.

Angle of Radiation

The fourth parameter effected by height is the angle of radiation or the angle of signal departure. Most amateurs are familiar with the concept of height versus radiation angle, having seen it illustrated using an image antenna that is hypothesized symmetrically below the air-earth interface. While image theory is a powerful tool in electrostatics, it can be misleading when used in antenna work because it implies a ground of infinite conductivity.\15 Fig. 4 presents a more realistic picture in which a dipole is located at point A; its axis is perpendicular to the page at a height H above the x axis, which represents lossy ground. In order to obtain maximum signal strength at remote point P, the direct ray (AD) and the ground reflected ray (BC) must arrive at point P in phase. Since earth reflection shifts the phase of a horizontally polarized wave by 180 degrees, the phase relationship of the two signals arriving at point P is controlled by the length of ray AB. When AB is an odd number of half wavelengths in length, the two components will add in phase and create a lobe. Conversely, when AB is an even number of half wavelengths long, a null will be created. Fig. 5 illustrates lobes at 30 degrees.

While plural lobes may occur at some heights, it is generally the lowest lobe that is of interest. Its location may be approximated, with good accuracy, by a simple formula in which /a is the angle of radiation, expressed in degrees above the horizontal, and H is the height of the dipole above ground.

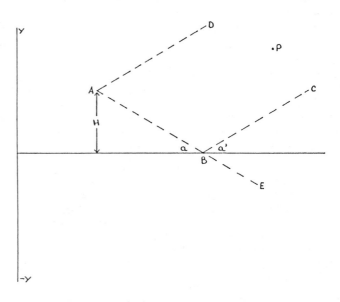

Fig. 4 -- Ray geometry of a dipole above lossy earth.

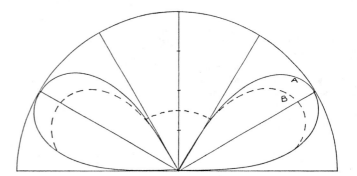

Fig. 5 -- The lobes of a dipole located one-half wavelength above a perfect conductor (solid line) and above lossy earth (dashed line).

$$\underline{/a} = \frac{14.33}{H} \quad \text{(H is in wavelengths)}$$

Our main concern herein is to determine what effect lossy ground has on the lobes. If the earth in Fig. 4 were a perfect electrical conductor, all of the energy in the incident ray AB would be reflected, and energy BC will equal that of AB. We know from geometric optics that the angle of incidence, /a in Fig. 4, will equal the angle of reflection, /a'. When the earth is not a perfect electrical conductor, part of the incident energy will be transmitted into the ground to be dissipated as heat, and the remainder reflected as ray BC. As before, /a equals /a'. The relationship:

Reflected Power
Incident Power

is known as the power reflection coefficient, often designated as $|\Gamma|^2$, and when multiplied by 100, indicates the percentage of incident power reflected. The formalism of this calculation is contained in my note at the end of this article. Table 1 illustrates the percentage of power reflected for various angles of incident and soil conditions. Note that the loss increases both as the angle of radiation (incidence) increases and as the soil quality decreases. Thus, the long-distance enthusiast who has raised his antenna to attain a favorable low angle of radiation will suffer less transmission loss than the amateur whose antenna is at a low height with its resulting high angle of radiation. In Fig. 5, the solid line indicates the lobes of a dipole one-half wavelength above a perfectly conducting ground, while the dashed curve is the same antenna above poor ground.\16 Note not only the difference in lobe amplitude, but also the fact that the complete null between the symmetrical lobes is partially filled when the earth is lossy.

Conclusion

The driving-point resistance of a horizontal dipole antenna is altered as ground proximity increases. Below about one fifth wavelength, the variation is controlled not only by height but also by soil characteristics. Both the bandwidth and the efficiency are affected when the near field comes into earth contact. While the bandwidth may be broadened in a desirable fashion, it is done at the expense of efficiency. High antennas, with their predominantly low angle of radiation, suffer less transmission loss over poor quality ground than do low antennas with their high angle of radiation.

Author's Note

When the Fresnel reflection coefficient, Rh, is squared, it will yield a power reflection coefficient. For horizontal polarization:

$$R_h = \frac{\sin\underline{/a} - \sqrt{\varepsilon' - \cos\underline{/a}}}{\sin\underline{/a} + \sqrt{\varepsilon' - \cos\underline{/a}}}$$

where ε' is the complex permittivity of the medium. Refs. 1 and 17 contain good discussions about Fresnel coefficients.

References

\1 Burrows & Atwood, **Radio Wave Propagation,** Academic Press, New York City, 1949.

\2 Terman, **Radio Engineers Handbook,** McGraw-Hill, New York, 1943.

\3 Fine, **An Effective Ground Conductivity Map for the Continental United States,** Proceedings of the IRE, September 1954.

\4 Jordan & Balmain, **Electromagnetic Waves & Radiating Systems,** Prentice-Hall, Englewood Cliffs, New Jersey, 1968.

\5 IEEE, **Recommended Guide for Measuring Ground Resistance and Potential Gradients in the Earth,** Std. 81-1962, IEEE, New York, 1962.

\6 Blake, **Antennas,** John Wiley & Sons, New York, 1966.

\7 Ramsey & Dreisbach, **Radiation and Induction,** Proceedings of the IRE, New York, 1966.

\8 Williams, **Antenna Theory and Design,** Vol. II, Sir Isaac Pitman, London, 1950.

\9 Schelkunoff & Friis, **Antenna Theory and Practice,** John Wiley & Sons, New York, 1952.

\10 Ma, **Theory and Application of Antenna Arrays,** John Wiley & Sons, New York, 1974.

\11 Sommerfield & Renner, **Radiation and Earth Absorbtion for Dipole Antennae,** Wireless Engineer, London, September 1942.

\12 Hansen, **The Radiation Efficiency of a Dipole Antenna Located Above an Imperfectly Conducting Ground,** IEEE Transactions, APS, New York, November 1972.

\13 IEEE, **Standard Definitions of Terms for Antennas,** IEEE, New York, 1983.

\14 Kraus, **Antennas,** McGraw-Hill, New York, 1950.

\15 Rao, **Elements of Engineering Electromagnetics,** Prentice-Hall, Englewood Cliffs, New Jersey, 1977.

\16 Ladner & Stoner, **Short Wave Wireless Communication,** Chapman & Hall, London, 1942.

\17 Radford (contributor), **The Handbook of Antenna Design,** Peter Peregrinius Ltd., London, 1983.

Additional Reading:

Bhattacharya, **Input Resistance of Dipoles Over Homogeneous Ground,** IEEE, Transactions, APS, New York, 1963.

Miller, Poggio, Burke & Selden, **Analysis of Wire Antennas in the Presence of a Conducting Half Space,** Part II, Canadian Journal of Physics, Vol. no. 50, Montreal, 1972.

Karwowski, **Low-Frequency Approach to the Problem of a Horizontal Wire Antenna Above an Imperfect Ground,** IEE, Proceedings, Vol. 131, Part H, London, 1984.

	$5°$	$10°$	$20°$	$40°$
Good soil	93%	87%	76%	60%
Average soil	92%	85%	73%	55%
Poor soil	89%	80%	64%	43%
Residential & Industrial	82%	68%	46%	25%

Table 1. Percentage of power reflected for various angles of radiation and soil conditions.

Biography

Robert B. Sandell began his Amateur Radio activity in 1937 and currently holds the call letters W9RXC. He served with the United States Navy in World War II working with the technical aspects of radar. He is a graduate of the University of Minnesota and has studied antenna and propagation theory at The Georgia Institute of Technology, George Washington University, the Southeastern Center for Electrical Engineering Education, and the University of Illinois.

He is a member of the Institute of Electrical and Electronic Engineers, the Antenna and Propagation Society, the Microwave Theory and Techniques Society, the Antenna Measurement Techniques Assoc., and is an affiliate member of the American Institute of Physics.

He was a participant at the 1981 and 1983 International Conference on Antennas and Propagation, sponsored by the Institution of Electrical Engineers, (London), at the University of York, England, and the University of East Anglia, England, respectively.

Antenna Polarization

Gerd Schrick,* WB8IFM

*4741 Harlou Drive, Dayton, Ohio 45432

In point-to-point communication using electromagnetic waves, the antennas are usually positioned and oriented in such a way as to maximize the transfer of energy. This requires pointing the antennas in the direction of maximum gain and using the proper polarization.

A problem arises when one or both of the stations are mobile and antenna rotation becomes impractical. Traditionally, this problem was solved by using vertical antennas. In aeronautical mobile operation, however, this problem is further aggravated by requiring elevated radiation. Finally, in space communication, the spacecraft moves overhead and rotates (spins) at the same time, so the question of polarization becomes a major concern. See Fig. 1.

Electromagnetic radiation is best understood by looking back at the basic radiator: a short loaded dipole (Fig. 2). By alternately charging the end spheres of this dipole with positive and negative electricity at a high rate, a propagating electromagnetic field is generated with preferred directions in regard to the orientation of the dipole. It should be understood here that positive means a lack of electrons and negative means a surplus.

The radiation of a dipole is the familiar doughnut shape directional pattern with omnidirectivity around the axis and no radiation off both ends of the dipole (the hole of the doughnut). Refer to Fig. 3. In this pattern the polarization of the electric field vector at some distance (far field) at the point of maximum radiation is in line with the axis of the dipole. As a general rule then, the polarization of an antenna is parallel to the current carrying elements. As one deviates from the maximum of the radiation pattern, the field vector, and therefore the optimum polarization of the receiving antenna, follow the doughnut pattern (Fig. 4). It is helpful to keep these simple facts in mind for the following discussion.

In practice, antennas for HF are usually horizontally polarized. This helps to minimize ground conductivity problems. Vertical antennas, in order to have the desirable low-angle radiation, require good ground conductivity or radials. On the higher frequency bands, where line-of-sight propagation prevails, the question of polarization is less important as long as it is compatible. Therefore, for 2-m FM, vertical polarization is used because it can compensate for moving objects. In the more serious DX work that takes place on VHF/UHF, horizontal polarization is used, taking advantage of the gain afforded by ground reflection.

Fig. 1 — Communication with moving platforms.

Fig. 2 — The short dipole (Hertz dipole).

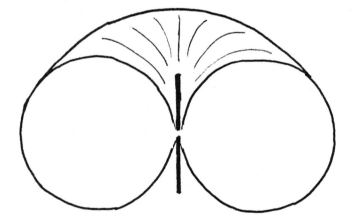

Fig. 3 — The familiar doughnut radiation pattern of a dipole.

But what about circular polarization? It is often used by commercial stations, the military and in space communication. In general, all single-fed antennas generate a linear-polarized wave such as the elementary dipole described above. Examples are depicted in Fig. 5. As can be seen, the shape of the antenna does not result in any special polarization other than linear. The outer lengths of the antenna conductors are merely used to resonate the antenna or provide proper phasing (as in the case of the quad) to other elements for the same antenna structure. **The RSGB VHF/UHF Manual** states that, "Being merely a dipole bent into circular shape, the popular halo antenna for 2 m is far from omnidirectional. Most of the radiation from a half-wave

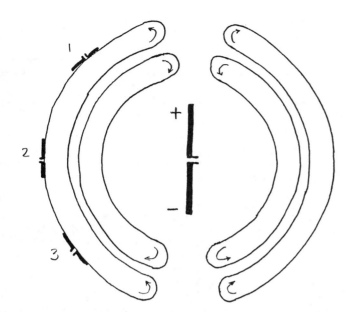

Fig. 4 -- Sample electrical field of a transmitting dipole and optimum alignment of receiving dipoles 1...3.

element is from the middle of the radiator, regardless of what is done with the ends."[1]

Now, when considering circular-polarization waves or antennas, contemplate one more dimension: **time.** It is easiest to visualize a circular-polarized wave by imagining a normal transmitting dipole rotating at the frequency of the wave. Take a 146-MHz signal for example. The dipole rotates 146 million times per second. If this dipole is opposed by a receiving dipole in any arbitrary linear fixed position, it would intercept part of the transmitted wave, but not all. As a matter of fact, it will receive 50% of the power that it could if both dipoles were perfectly aligned. On the other hand, if both dipoles are fixed and poorly aligned, or even at a right angle, less than 50% or nothing at all would be received. Finally, if the receiving antenna is capable of receiving the rotating wave, power transfer would again be 100%. This is shown in Fig. 6 and Box 1.

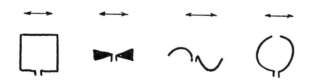

Fig. 5 -- Different shaped antennas generate the same linear polarization.

Polarization mismatch is a term used to describe the less than 100% power transfer. Circular polarization then is the best choice when dealing with unknown or changing polarization. The event of the first ham in space (the Columbia space shuttle mission no. 9, Dec. 1983) is one example of how Amateur Radio operators showed concern about using proper polarization. The interaction of various polarizations is summarized in Fig. 7.

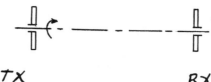

Fig. 6 -- The rotating transmitter dipole vs. the stationary receiving dipole. (Effects of polarization mismatch.)

TX	RX	
non rotating	aligned	100%
rotating	any position	50%
non rotating	poorly aligned	<50%
non rotating	at right angle	0

It is clearly impossible to rotate the antenna physically. So what are the methods to generate the circular polarized wave? There are two methods known and used. The first consists of feeding two or more independent linear-polarized antennas so as to generate a circular-polarized wave. The second method uses a wave-structure antenna where the electric wave generated is guided along in a circular way. Both methods are widely used, but the second method is used mostly for the higher frequencies with the 70-cm band forming the dividing line.

So far we have explained the function of these antennas in terms of transmitting. These same antennas, of course, function as well in a similar manner as receiving antennas.

TX \ RX	—•—	⫯	↻	↺
—•—	0	∞	3	3
⫯	∞	0	3	3
↻	3	3	0	∞
↺	3	3	∞	0

Fig. 7 -- Interaction of various polarizations (extra attenuation in dB).

Before going into details about the actual antennas used, a few more terms should be explained. Two rotational directions are possible with the circular-polarized antenna. They are labeled right hand circular polarized (RHCP) and left hand circular polarized (LHCP). By convention, the polarization is viewed from the transmitting antenna in the direction of the radiation (Fig. 8A). Elliptical polarization is a term often heard. It refers to a two-feed system antenna and means that one of the

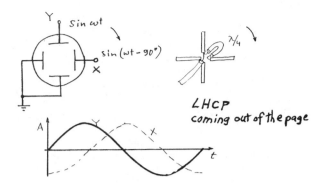

Fig. 8 -- The generation of a circular polarized wave (Lissajous figure on the scope, cross dipole).

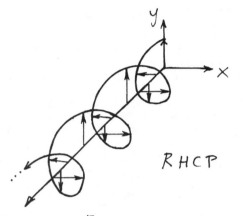

Fig. 8A -- Right hand circular-polarized wave; electric field vector.

field vectors generated is stronger than the other, leading to a deformation of the circle into an ellipse. Most practical antennas exhibit some amount of ellipticity. If more than two antennas are used, the resulting pattern could be distorted more from the circle than a simple ellipse. Such is the case with the 144-MHz downlink signal of OSCAR 10; apparently one of the three antennas, excited for RHCP, was damaged when it was deployed.

Crossed Dipoles/Yagis

To generate a circular-polarized wave, which we have described as a time varying linear-polarized wave, two crossed dipoles are fed with the same signal, but out of phase. The phase difference determines the polarization, as does the orientation of the dipoles. See Box 2. The two dipoles are at right angles to each other and in the same plane, and the signal to one of the dipoles is delayed by a quarter wave. The way this works can be compared to generating a circle on an oscilloscope (Lissajous figures). If the same sine wave is fed to the x and y plates of the scope, a diagonal line is obtained. If we now delay (or shift the phase of) the signal to one of the plates, an ellipse will be formed. When the delay is right, a circle is formed as in Fig. 8.

What can be performed with the dipoles can also be done with Yagi antennas. Sometimes the antenna may be shifted on the boom by a quarter wavelength,

in which case the feed lines to the radiators should be equal. By adding a half wave to the quarter wave delay or to one of the equal lines, the sense of rotation is reversed. Switching and inserting lengths of coaxial cable will switch the antenna to a number of linear polarizations as well as to LHCP or RHCP. Note an interesting effect: From the back of the antenna a reverse circular-polarized wave attenuated by the front-to-back ratio eminates. These features are, however, difficult to implement at UHF and higher. Fig 9. suggests some feed-line and matching approaches depending on the quarter-wave transformer in use.

Helical Antenna

Because of its physical size, the axial helical beam is practical from about 435 MHz and up.[2] It uses only one feed and generates a circular polarization in space by guiding the wave along the helical structure. Here it is easy to determine the rotational sense just by looking at the helix. The helical antenna has one polarization and cannot be switched. The antenna design is not critical and is rather broadband. Bandwidth ratios of 2:1 are not uncommon. This feature makes it possible to use the same helix on 144 and 220, or 220 and 435 MHz. Fig. 10, courtesy of **RSGB Handbook**, offers some basic dimension and gain figures.[3]

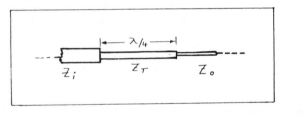

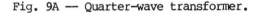

Impedance Equation: $\dfrac{Z_T}{Z_i} = \dfrac{Z_o}{Z_T}$ or $Z_T = \sqrt{Z_i \times Z_o}$

Fig. 9A -- Quarter-wave transformer.

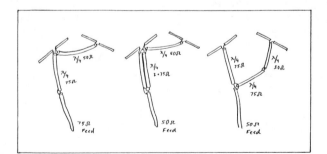

Fig. 9B -- Methods of feeding crossed dipole/yagis for circular polarization using quarterwave delay lines and transformers.

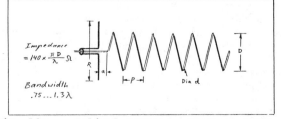

Fig. 10A -- Helical antenna, courtesy RSGB.

OSCAR 10

Refer to Fig. 10. Aside from the fact that the 2-m-gain antenna structure (Fig. 11) has been damaged, the elliptical orbit naturally varies the distance to earth. However, the attitude relative to space remains the same, and the radiation pattern is designed to keep the signal strength constant. This does not take polarization into consideration, and many observers with the capability to switch the polarization type have found it superior for different OSCAR 10 positions. At or near apogee (the highest position), the intended RHCP is definitely optimum. The exclusive use of this polarization, although not always optimum, is definitely sufficient for operation.

Fig. 12 shows the OSCAR 10 beacon signal at 145.810 MHz recorded by Guenter Schwarz, DL1BU, using a switchable crossed Yagi arrangement which permitted four linear polarizations and RHCP and LHCP. The recording took place at apogee and is most indicative of the intended and achieved link energy transfer. The rapid fluctuations of the signal (referred to as spin fading or spin modulation) most noticeable on all the linear polarizations is caused by the damaged antenna in conjunction with the spinning (about 2 revolutions per second) of the satellite. The fact that the circular-polarized antenna does not pick up this "spin fading" does reveal that the real cause is related to polarization mismatch. The circular-polarized antenna (RHCP) takes care of this effect nicely.

The LHCP is received at about 12 dB weaker and is akin to the front-to-back ratio. The linear polarizations are all expected to produce a 3-dB weaker signal which is generally confirmed by the recordings.

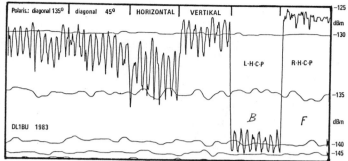

Fig. 12 — Plot of the OSCAR 10 beacon signal.

Box 1

Gain of Circular Polarized Antennas

To consider the gain of a circular-polarized antenna, it is best to consider the crossed Yagis. If the Yagis were side by side or stacked, the gain of the arrangement would obviously be 3 dB higher than a single Yagi. In the cross arrangement, this 3 dB is not available and is considered polarization loss. However, if the intercepted wave were circular polarized, there would be no polarization loss for the crossed Yagi and 3-dB polarization loss for the side-by-side Yagi, thus making the two antenna arrangements equally efficient.

Box 2

How to Determine Polarization of the Crossed Dipole/Yagi

Locate the dipole halves or gamma match rods that are connected to the center conductor of the coaxial lines. Rotate the tip of the dipole half directly connected to the feed line on the short way to the tip of the dipole half connected through the quarter wave delay line. You now have the sense of rotation. If you then move out into the direction of the radiation (forward), the screw formed will either be right hand or left hand corresponding to RHCP or LHCP. If you move to the opposite side of the antenna (backwards), the polarization is opposite. Some delay lines are 3/4 wavelength; in that case the polarization is also opposite (reverse) from the one at 1/4 wavelength.

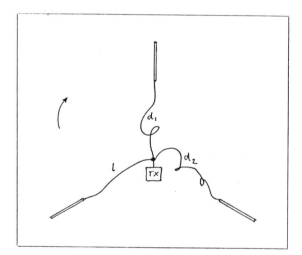

Fig. 11 — OSCAR 10, 2-m antenna feed system for circular polarization. Lengths of coaxial feeds: d1 = ℓ + 1/3 λ ; d2 = ℓ + 2/3 λ .

References

\1 Jessop, G. R., **VHF/UHF Manual**, RSGB, 1971.
\2 Kraus, John D., **Antennas**, McGraw-Hill, Inc., New York, 1950, Chapter 7, p. 173.
\3 Radio Society of Great Britain, **RSGB Handbook**, 4th edition, 3rd printing, Feb. 1971, pp. 14-20 to 14-21, 9-28.

Final Comments

Space communication is the wave of the future. The great capacity of the line-of-sight waves used, as well as their low distortion, have already proved its superiority to the HF band. For example, worldwide television could hardly be realized by HF. Hams must become more involved in the high frequencies. Understanding wave propagation and polarization is part of it.

Table 1

The helical antenna — dimensions and performance.   Courtesy of RSGB.

Band	Dimensions				
	D	R	P	a	d
General	0.32 λ	0.8 λ	0.22 λ	0.12 λ	
432	8-3/4"	22"	6"	3"	1/4"
1250	3"	7"	2"	1-1/8"	1/4 to 1/8"
Turns	6	8	10	12	20
Gain	12 dB	14 dB	15 dB	16 dB	17 dB
Beamwidth	47°	41°	36°	31°	24°

Baluns: What They Do And How They Do It

By Roy W. Lewallen,* W7EL

*5470 SW 152nd Ave., Beaverton, OR 97007

I've always been a bit bothered by baluns, since I was never sure what they are supposed to do, let alone how they might go about doing it. The majority of articles deal with various ways of building and testing baluns, or the advantages of one type over another, but almost never a word about when or why a balun is necessary, if at all. Like most amateurs, there have been few occasions when I have been able to tell if a balun has any effect on an antenna system, and when it has, the effect hasn't always been good! The turning point came when I was trying to measure the resonant frequency of a folded dipole through a one-wavelength coaxial line. The bridge null varied a great deal as I moved my hand around the coaxial cable, or if the line or bridge was moved. A hastily constructed balun installed at the center of the dipole eliminated the problem. But why?

I found a brief, but clear explanation of one phenomenon involved in a paper by Maxwell, W2DU.\1 However, many questions remained. This led me to an investigation of just how baluns are supposed to work, and what problems they are supposed to cure. One surprising conclusion I found from my research is that one popular type of balun, when properly designed and used in an antenna system, may not solve the problems that baluns are expected to solve. Other results indicated that the type of feed line (balanced or unbalanced) has little to do with how well a **system** is balanced. In order to verify, or refute, the theoretical results, several experiments were carefully set up and run, and the data analyzed. The result is a much clearer view of the operation of baluns in antenna systems, and some definite "dos" and "don'ts" regarding their use.

What Problems are Baluns Supposed to Solve?

Baluns usually solve problems caused by an imbalance. An imbalance of what? To answer this question, we need to look at current flow in transmission lines.

In a coaxial cable, the currents on the inner conductor and the inside of the shield are equal and opposite. This is because the fields from the two currents are confined to the same space.\2 With the presence of skin effect, a different current flows on the outside of the shield than on the inside.\3 The current on the outside, if significant, causes the feed line to act like an antenna, radiating a field that is proportional to this current.

A twin-lead feed line has similar properties, despite its different physical nature. Since it is physically symmetrical, if the currents flowing through the conductors are equal and opposite, the radiation from the line is minimal (assuming that the conductor spacing is very small relative to a wavelength). However, several factors may cause the

currents in the two conductors to be imbalanced, that is, other than equal and opposite. If this happens, the balanced feed line will radiate like a coaxial cable that has current on the outside of the shield. This occurs because the components of the currents on the two conductors that are equal and opposite create fields which cancel. But the field from any remaining component on either conductor (called a common-mode, secondary-mode, or antenna current) will cause radiation.\4,5,6 In this article, the current on the outside of the coaxial shield, or the antenna current on the twin lead, will be called the imbalance current: They are caused by the same things and produce the same effects.

Imbalance current, on either kind of line, is the cause of a number of undesirable effects:
o pattern distortion (caused by the feed-line radiation adding to the antenna-radiated field, or by unequal currents in the antenna halves)
o TVI (radiation from a feed line coupling into nearby television sets, house wiring, and so on)
o RF in the shack (caused by a "hot" radiator — the feed line — residing in the shack)
If you have read other articles on baluns, you'll recognize these as the problems baluns are supposed to solve. What isn't usually too clear is that they are all caused by **current imbalance, on either coaxial or twin-lead feed line.** Of course, if the imbalance current is sufficiently small to begin with, a balun is not necessary at all. Or it can be said that a properly designed balun will not solve the problem being experienced.

What Causes System Imbalance?

The first cause of imbalance currents was explained by Maxwell. It will be repeated here for completeness. When a balanced antenna is fed with coaxial cable (Fig. 1), the outside of the shield appears as an extra, separate conductor connected to the right side of the antenna at the feed point. The current in the cable's center conductor flows into the left half of the dipole. The equal and opposite current on the inside of the shield flows partly into the right half of the dipole, and partly along the outside of the shield. The proportion of current which flows each way is determined by the relative impedances of the two paths. The current on the outside is the greatest when the total effective length of the path along the outside of the coaxial cable from the antenna to ground is an integral number of half wavelengths, since this makes the impedance presented by the undesirable path relatively low. If the rig is effectively an odd number of quarter wavelengths from actual ground, it is at a voltage maximum and can be hot. On the other hand, there are other combinations of lengths for which the imbalance current will be negligible — cases where a balun does not make any noticeable difference.

The obvious solution to this problem is to feed

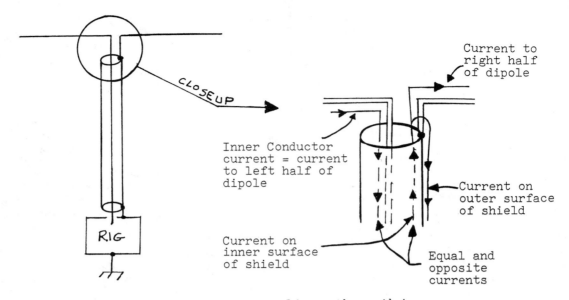

Fig. 1 — Imbalance caused by another path to ground from only one side of the dipole.

the balanced antenna with a balanced feed line — twin lead. This solves the problem neatly, until you encounter the problem which most of us have today, illustrated in Fig. 2. Suppose that we went ahead and connected the line as shown schematically in Fig. 3. If the rig could be totally isolated from ground, the feed-line conductor currents would be equal and opposite, just as they would be if coaxial cable were used; the imbalance current would be zero, and the feed line would not radiate.\7 However, when we connect the rig to ground, as shown by the broken line, we've again provided a third conductor in parallel with the right side of the feed line, and the same problem occurs as with the coaxial cable (see Fig. 4). So either type of line is unbalanced if a direct path to ground is provided from one side, and both can be a balanced, non-radiating line if the imbalance current is eliminated.

Imbalance current can be caused also by situations where the two sides of the antenna are not precisely symmetrical: Coupling to nearby objects, the tilt relative to ground, or slight differences in lengths of the two antenna halves.\8 Another cause of imbalance currents is induction. If the feed line is not exactly placed at a right angle to the antenna, a net current is induced into it by the antenna field. This current appears as an imbalance current. At UHF, where the diameter of coaxial cable is a substantial fraction of the length of the antenna elements, coaxial line is more difficult to place symmetrically relative to the antenna than twin lead is (this is sometimes given as the only reason for using a balun!). The problem is negligible at VHF (except perhaps with very large diameter coaxial cable) or below.

What Baluns Do

Let's recall what we want a balun to do: cause the currents in the feed-line conductors to be equal in magnitude and opposite in phase, resulting in a zero imbalance current. How well do the popular balun types do this?

One type of balun is known as a transformer-type balun or balun with a tertiary winding (Fig.

5).\9 This type is commonly used for providing single-ended to differential conversion for driving balanced mixers, push-pull amplifiers, and so on. It seems to be suitable for our purpose. An analysis of its operation (see Appendix 1) shows that it does indeed perform an unbalanced-to-balanced conversion. The voltages at the balanced port are caused to be equal, and opposite, in phase relative to the cold side of the unbalanced port. Thus, the use of this sort of balun will eliminate the problem of current flow on the outside of a line **only if the antenna is perfectly balanced.** There is nothing gained by forcing the voltages of the two antenna halves, whether balanced or not, to be equal and opposite relative to the cold side of the balun input (usually connected to the shield of a coaxial feed line), since the antenna field is proportional to the **currents** in the elements, not the voltages at the feed point. I will call this type of balun a voltage balun to emphasize that it balances the output **voltages** regardless of load impedances.

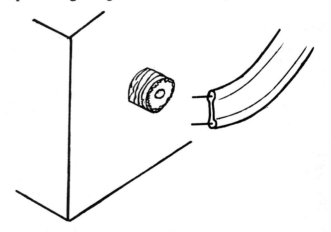

Fig. 2 — One solution to the imbalance problem is to feed the balanced antenna with a balanced feed line — twin lead.

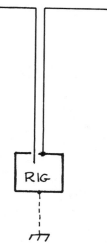

Fig. 3 -- If the rig were isolated from ground, the feed-line conductor currents would be equal and opposite, the imbalance current would be zero and the feed line would not radiate.

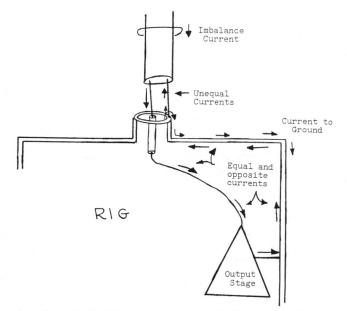

Fig. 4 -- When the rig is connected to ground, a third conductor in parallel with the right side of the feed line is introduced. This causes an imbalance in both coaxial cable and twin lead feed lines.

Another type of balun that appears in the literature has been called a choke-type balun (Fig. 6).\10 It resembles the voltage balun, except that the tertiary winding is missing. The analysis of both types of balun in Appendix 1 shows that the effect of a tertiary winding is not a minor one. The two types of baluns produce fundamentally different results. The voltage balun causes equal and opposite **voltages** to appear at the balanced port regardless of load impedances, but the second type of balun causes equal and opposite **currents** on the conductors at both ports for any load impedances. For this reason, I will call this type of balun a current balun. Intuitively, the current balun produces the sort of effect we would expect. When wound with twisted pair or twin lead, it is nothing but a bifilar RF choke that impedes any net current which tries to flow through it. When wound with coaxial

cable, it can be visualized as an RF choke acting only on the outside of the coaxial-cable shield, reducing the current to a very small value. This is the exact function a balun needs to accomplish when used in an antenna system.

A current balun can be constructed by winding coaxial feed line into a coil, winding either type of feed line onto a core, or by stringing ferrite cores along either type of line.\11 Even if the balun is mediocre, there will be no effect on the desired properties of the line itself (impedance, electrical length, SWR, and so on). A less-than-perfect voltage balun can have a profound effect on the impedance seen at its input because of the tertiary winding. Impedance-transforming (4:1) baluns are discussed in Appendix 3.

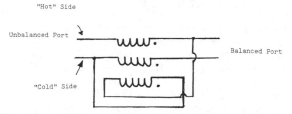

Fig. 5 -- Voltage balun (see text). The bottom winding is sometimes referred to as a tertiary winding. All windings are closely coupled.

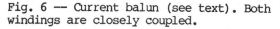

Fig. 6 -- Current balun (see text). Both windings are closely coupled.

Experiments

A series of experiments was designed to test the validity of the results of the theoretical investigation. A 10-meter dipole was set about 12 feet above the ground, and about five feet above the edge of an elevated wooden deck (Fig. 7). One-half wavelength from the center of the dipole, a 4-foot rod was driven into the ground, which was completely saturated with water at the time the experiments were run (during November, in Oregon). To further lower ground-system impedance, six radials were placed on the ground around the ground rod. Two feed lines were cut to a half wavelength: one of RG-59/U coaxial cable, and one of 72-ohm transmitting twin lead. The velocity factors of the cables were not taken into account, since the intent was to have the outside of the coaxial cable, or the two parallel conductors of the twin lead, be an electrical half-wavelength long. A low-power 10-meter transmitter located at, and connected to, the ground system was used as a signal source.

Current probes and baluns were built as described in Appendix 2. Two of the current probes were permanently wired into each side of the dipole near the feed point, and a third was used for all feed-line measurements. A single detector was used for all measurements, and it was calibrated over the range of encountered output levels by using a signal source and precision attenuator. The results of the experiment have been corrected to account for the measured nonlinearity of the detector.

No attempt was made to keep the power level or impedance match constant from one test to another. When running an experiment with no balun, a current balun, and a voltage balun, the only variation in the system was to change the balun. Initially, the intent was to use the antenna current probe readings as a measure of current balance in the antenna halves. However, a case was encountered in which the antenna halves showed equal currents, but a large imbalance current was measured in the feed line at the antenna feed point — a seemingly impossible combination! (The equal antenna currents were even more suspicious because no balun was being used, and the antenna had intentionally been made nonsymmetrical for that test.) A bit of thought provided the answer. The imbalance current is measured by placing the feed line through the current-probe toroid. In conjunction with the detector, it measures the magnitude of the vector sum of all currents flowing through the toroid. Each antenna current probe, with the detector, measures the magnitude of the current in each half of the antenna, at the feed point. What must be happening is that the currents in the dipole halves are equal in magnitude, but not 180 degrees out of phase. A check of the current-probe outputs with a good-quality dual-channel oscilloscope confirmed the hypothesis: The currents were 230 degrees, rather than 180 degrees, apart, although equal in magnitude. What an interesting pattern that dipole would have! But this illustrates how misleading the magnitudes of element currents can be when judging balance. Measuring the imbalance current in the feed line at the feed point does, however, provide a good indication of the balance of the currents in the antenna halves. If the imbalance current is very small, the currents in the sides of the antenna must be nearly equal in magnitude **and** opposite in phase. A significant imbalance current, on the other hand, indicates that one or both conditions have not been met.

Measurement of the imbalance current on the feed line also indicates how much the feed line will radiate. The imbalance current at the rig provides a measure of RF in the shack. In the following tests, the magnitude of the current was measured in each conductor, then the magnitude of the imbalance current was measured by placing the complete feed line through the current-probe toroid. A single figure of merit, **balance**, was calculated as:

$$\text{balance (dB)} = 20 \log \left[\frac{\text{(average of magnitudes of currents in each conductor)}}{\text{(magnitude of imbalance current)}} \right].$$

Experiments 1 through 4 were done using a nominally symmetrical dipole, although results indicate that some asymmetry was present. For experiments 5 through 7, the dipole was intentionally made nonsymmetrical by lengthening one side by five inches, and shortening the other side by the same amount.

Experiments 1 and 5:

See Fig. 8. The dipole was symmetrical for experiment 1, nonsymmetrical for experiment 5.

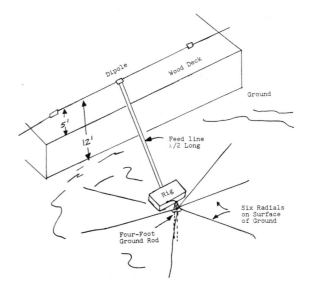

Fig. 7 — Experimental setup.

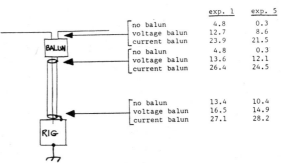

	exp. 1	exp. 5
no balun	4.8	0.3
voltage balun	12.7	8.6
current balun	23.9	21.5
no balun	4.8	0.3
voltage balun	13.6	12.1
current balun	26.4	24.5
no balun	13.4	10.4
voltage balun	16.5	14.9
current balun	27.1	28.2

Fig. 8 — Setup and results of experiments 1 and 5. Numbers are measured balance in dB.

Discussion

If the dipole balance (symmetry) were indeed perfect for experiment 1, we would expect the currents in the sides of the dipole to be unbalanced, resulting in imbalance current on the feed line. This is because the outside of the coaxial shield appears as a conductor in parallel with half of the dipole. Also, either a current or voltage balun should reduce the imbalance current to zero. Since the feed line is placed symmetrically relative to the antenna, no additional current should be induced into the feed line, so the imbalance should also be quite small at the rig end of the line when either type of balun is used.

With the nonsymmetrical dipole (experiment 5), we would expect the voltage balun to do worse than in experiment 1. We would also expect the current balun to do about the same, and the no-balun case to be considerably worse.

Results

In experiment 1, the voltage balun did not perform as well as the current balun, indicating some asymmetry in the dipole. At the frequency chosen, the small differences in connections and a slight tilt of the antenna could easily account for what happened. When no balun is used, a curious result is the much better balance at the rig end than at the antenna end of the feed line. This may be because the feed lines weren't exactly an effective half wavelength long, because there was a wire of about six inches in length connecting the rig to the ground system, or because the feed line was doubled back on itself for a short distance near the rig to provide strain relief. Perhaps the doubling back generated enough inductance to cause a current balun, or RF choking effect. The better balance at the rig end can be seen in the results of all experiments.

The no-balun result was worse with the non-symmetrical dipole than the symmetrical one, as expected, and the current balun did about the same in both cases. The voltage balun, although slightly worse with the nonsymmetrical antenna, was better than expected, but still definitely inferior to the current balun.

Experiments 2 and 6:

See Fig. 9. The dipole was symmetrical for experiment 2, nonsymmetrical for experiment 6.

Discussion

The results of these experiments should duplicate those of the previous pair, since the feed line is placed symmetrically relative to the antenna to avoid induced current. The only difference is that the balun is placed farther down in a symmetrical or nonsymmetrical system.

Results

The trend is clearly the same as in experiments 1 and 5; the current balun provides the best balance, the voltage balun is second best, and a feed line with no balun is the worst case. The balance with no balun was better in this experiment, however, (except at the rig end with the non-symmetrical antenna, which was about the same), and the balance at the rig end was substantially better when using the current balun. Time did not permit me to run additional experiments to explain these differences, but the ability of the current balun to achieve superior balance was again illustrated.

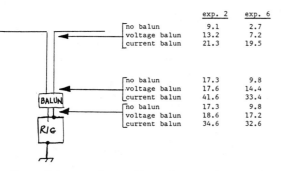

	exp. 2	exp. 6
no balun	9.1	2.7
voltage balun	13.2	7.2
current balun	21.3	19.5
no balun	17.3	9.8
voltage balun	17.6	14.4
current balun	41.6	33.4
no balun	17.3	9.8
voltage balun	18.6	17.2
current balun	34.6	32.6

Fig. 9 — Setup and results of experiments 2 and 6. Numbers are measured balanced in dB.

Experiments 3 and 7:

See Fig. 10. The dipole was symmetrical for experiment 3, nonsymmetrical for experiment 7. The voltage balun was connected with the balanced port toward the antenna.

Discussion:

These experiments, and experiment 4, were conducted to test the idea that coaxial cable and twin-lead feed lines would behave in the same fashion, as theorized earlier. If so, the results of

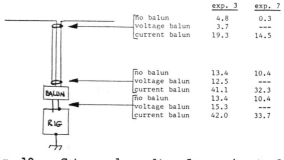

	exp. 3	exp. 7
no balun	4.8	0.3
voltage balun	3.7	---
current balun	19.3	14.5
no balun	13.4	10.4
voltage balun	12.5	---
current balun	41.1	32.3
no balun	13.4	10.4
voltage balun	15.3	---
current balun	42.0	33.7

Fig. 10 — Setup and results of experiments 3 and 7. Numbers are measured balance in dB.

these experiments should be similar to those of the previous pair.

Results:

With no balun, the results were those of experiments 1 and 5 (the test with no balun was not rerun). With the current balun, the results were similar to those of experiments 2 and 6, indicating that coaxial cable can be used as a balanced feed line (in the sense discussed earlier) with a balanced or somewhat-unbalanced load. This data also points to the possibility that a current balun could be added to an existing antenna system at the rig end of the line, with results similar to those obtained by placing it at the antenna, in some cases at least. This would certainly be worth a try in systems where the symptoms indicate the need for a balun, but the antenna itself is difficult to get to. With the symmetrical antenna, the voltage balun made balance **worse** at both ends of the feed line than no balun at all. The balanced port of the voltage balun sees two unequal impedances to ground: the coaxial center conductor, ending in one dipole half, and the coaxial shield terminating in the other. The shield is capable of radiating but the inner conductor isn't, and the two are of different diameters, accounting for the different impedances. The voltage balun predictably generates unequal currents in the different impedances, causing additional current imbalance. A voltage balun was not evaluated in this application with a non-symmetrical dipole, having shown distinctly inferior results even with a symmetrical one.

Experiment 4:

See Fig. 11. The dipole was symmetrical. The voltage balun was connected with the balanced port toward the antenna.

Discussion:

Like experiments 3 and 7, this was intended to test the similarity between performance of the two kinds of feed line, provided that all other conditions and connections are the same. If the two feed lines act the same, the results should duplicate those of experiment 1.

Results:

The current balun again causes the predicted results, except it shows improved performance at the rig over experiment 1. In contrast, the voltage balun gave strikingly poorer balance at the antenna, and markedly poorer performance at the antenna end of the feed line, compared to no balun at all. If the antenna were completely symmetrical, there should be no current imbalance at the input end of

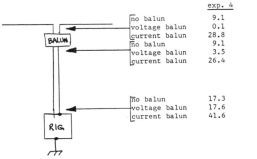

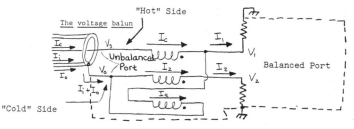

Fig. 11 — Setup and results of experiment 4.
Numbers are measured balance in dB.

	exp. 4
no balun	9.1
voltage balun	0.1
current balun	28.8
no balun	9.1
voltage balun	3.5
current balun	26.4
no balun	17.3
voltage balun	17.6
current balun	41.6

the voltage balun, but with the moderate (unintentional) imbalance presented by the actual antenna, the current balance on the feed line was seriously degraded. This configuration isn't likely to be used in actual practice, but helps illustrate the operation of the baluns and feed lines.

Conclusions

Although some aspects of the experimental results remain to be explained (as they always will be unless performed under extremely controlled conditions), they certainly support the theoretical analysis. The current balun gave superior balance at every measured point in each experiment. The voltage balun improved balance in most cases, explaining its acceptance in spite of the theoretically and experimentally demonstrated superiority of the current balun to cure the problems we have discussed.

As always, finding the answers to questions generates yet more questions. Lack of time did not permit experiments with the feed line placed non-symmetrically with respect to the antenna, to induce imbalance current into the feed line. The results of such an experiment should be interesting and enlightening.

Is there an optimum point in the feed line to place a balun? Suppose the effective distance along the feed line/ground wire from the antenna to ground is an integral number of half wavelengths, and the balun is placed a quarter wavelength below the antenna, as sometimes recommended. Wouldn't the imbalance current be conducted as before? Would induced current, if present, be reduced? What's the effect of poor coaxial shield coverage?

More work needs to be done in evaluating the various styles of current baluns (such as coaxial cable wound into a choke, coaxial cable wound on a ferrite or powdered iron core, insertion of the feed line through one or more ferrite cores, and so on) for their **primary** characteristic: causing currents to be equal in magnitude and out of phase. The method I've used is briefly described in Appendix 2, but how good is good enough?

The basic investigation reported here does answer some of the major questions regarding baluns. I now know what symptoms I can expect a balun to cure, why it will (or won't!) cure them, how to predict and measure the balun's success in doing so, and what type of balun to use. I hope you do, too!

Appendix 1: A Brief Analysis of Balun Operation

Analysis of both balun types assumes "ideal" operation: All flux is linked to all windings (coefficient of coupling is one), and each winding has sufficient self-impedance to make the magnetizing current negligible.

The Voltage Balun

Because of transformer action, $V1 - V3 = V2 - V\emptyset = V\emptyset - V1$. The third term comes about because of the "tertiary" winding shown at the bottom. Rearranging the last two terms, $V2 - V\emptyset = -(V1 - V\emptyset)$. So relative to $V\emptyset$ (the voltage at the cold side of the unbalanced port), the voltages at the balanced port are equal and opposite.

The current It flowing in the tertiary winding is, by inspection, $-(Ii + I\emptyset - I2)$ and also $(Ic - I1)$, so $-(Ii + I\emptyset - I2) = (Ic - I1)$. Because of the property of coaxial cable discussed in the body of the article, $Ii = -Ic$, so $Ic - I\emptyset + I2 = Ic - I1$; thus $I\emptyset = I1 + I2$. So for the current on the outside of the shield, $I\emptyset$, to be zero, load currents $I1$ and $I2$ must be equal and opposite. Since $V1$ and $V2$ are forced to be equal and opposite relative to $V\emptyset$, the only way for $I1$ and $I2$ to fulfill this requirement is for the impedances from each side of the balanced port to the cold side of the unbalanced port to be equal. Thus, only a perfectly balanced load will cause no current on the outside of the coaxial cable. Ironically, if this does occur, current $It = \emptyset$, and the tertiary winding accomplishes no function.

The Current Balun

In an ideal transformer of two windings having an equal number of turns, the currents in the windings are forced to be equal and opposite. So $Ic = -(Ii + Io)$. Again, $Ii = -Ic$, so $Ic = Ic - Io$, resulting in $Io = \emptyset$. This result is independent of the load impedances. And, since the load currents are the winding currents, they are also equal in magnitude and opposite in phase.

If the balun is constructed by winding coaxial cable on a core or into an air-core coil, or by stringing ferrite beads on the outside, the operation can be understood by observing that the inside of the coaxial cable "can't tell" what's going on outside. The currents on the inside — equal and opposite — happen regardless of the outside environment, but the construction causes a high impedance to current flow on the outside, acting like a choke to the imbalance current (hence the appropriate name choke balun). When constructed of twisted-pair line, the effect on imbalance current is the same and for the same reasons, but operation is more difficult to visualize.

The current balun

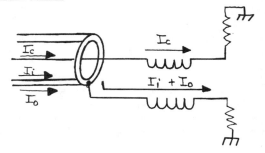

Appendix 2: Construction and Test of Baluns, Current Probes and Detector

Voltage Balun

The voltage balun was constructed using the method described in Ref. 8. A piece of no. 26 wire was laid along a length of RG-178/U cable (small-diameter Teflon-insulated coaxial cable), and heat-shrinkable tubing was applied over the assembly. The modified cable was wound on an FT82-61 core using ten turns. This construction method was decided on after trying to wind a balun with two pieces of coaxial cable in bifilar fashion, the shield of the second being connected as the tertiary winding. The latter construction method was much poorer in providing good voltage balance.

Voltage balance was evaluated by connecting the cold side of the unbalanced port to a ground plane and the balanced port to two resistors of unequal value, the other ends of which were connected to the same point on the ground plane. Using resistors of 27 and 54 ohms, the ratios of voltages appearing at the two resistors were measured as about 3/4 and 1-1/2 dB, depending on which resistor was connected to which lead of the balanced output.

Current Balun

The current balun consisted of 15 turns of RG-178/U coaxial cable on an FT82-61 core. Performance was evaluated by connecting the output end to 27- and 54-ohm resistors to ground, and measuring the voltages across them. A properly working current balun should generate twice the voltage across the 54-ohm resistor than across the 27-ohm resistor, regardless of which lead is connected to each resistor. The results were within 0.2 dB of theoretical, with either lead connected to either resistor.

Current Probe

The current probes were constructed as shown in Fig. 12. The output voltage equals ten times the current, in amperes, being measured. Insertion resistance is one ohm.

Detector

The detector is shown in Fig. 13. It was calibrated using a signal source and precision attenuators, at the operating frequency. Calibration using a dc source was found to be inaccurate.

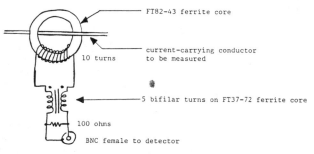

Fig. 12 — Current probe.

Fig. 13 — Detector.

Appendix 3: Impedance-Transforming (4:1) Baluns

The common 4:1 balun, shown schematically in Fig. A3-1, is a voltage balun. If used with a current balun as in Fig. A3-2, the combination acts like a 4:1 current balun. Or it can be converted to a 4:1 transforming current balun by adding a third winding, as shown in Fig. A3-3. A 1:1 voltage balun could be converted to a 4:1 current balun by reconnecting the existing windings. The difficulty with using this configuration is that, like the 1:1 voltage balun, all windings must be closely coupled, and rather severe impedance changes can occur because of transformer imperfections.

A better approach is shown in Fig. A3-4. Old-timers will recognize this as the configuration used by the balun coils commonly used some years ago. This balun does force equal and opposite currents at the input and output, so it is a true current balun, and it performs a 4:1 impedance transformation. Although it does require two cores which must **not** be coupled,* it has several advantages: It's much easier to tightly couple two conductors than three, it's much more forgiving than the other configurations, and it lends itself to easy construction. One method is simply to wind coaxial cable on two cores, with the center conductors being the conductors shown on the outsides in the figure. This balun can also be used in all-coaxial-cable systems. Besides effecting a 4:1 impedance transformation, it will greatly reduce any current flowing on the outsides of the lines.

* If ferrite rods or air-core coils are used, don't place them end to end. Place them side by side and spaced a fair distance, or, better yet, at right angles. Less care needs to be taken with toroidal coils.

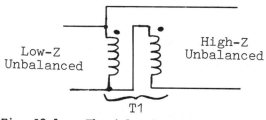

Fig. A3-1 — The 4:1 voltage balun.

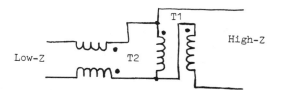

Fig. A3-2 — The 4:1 voltage balun used with a 1:1 current balun.

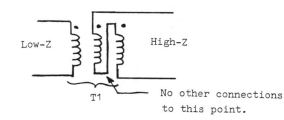

Fig. A3-3 — A 4:1 current balun.

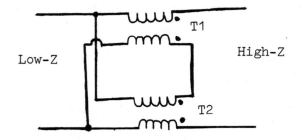

Fig. A3-4 — A superior 4:1 current balun.

References

\1 Maxwell, Walter, W2DU, "Some Aspects of the Balun Problem," **QST**, March 1983, p. 38.

\2 If a perfect shield is assumed (a reasonable approximation for this analysis), the result follows directly from Ampere's Law. For a more detailed explanation, see **Electromagnetic Energy Transmission and Radiation**, by Richard B. Adler, Lan Jen Chu, and Robert M. Fano (Wiley, 1960).

\3 A very clear development of the phenomenon of skin effect may be found in Chapter 7 of **Electric Transmission Lines** by Hugh H. Skilling (McGraw-Hill, 1951).

\4 Winningstad, C. Norman, "Nanosecond Pulse Transformers," **IRE Transactions on Nuclear Science**, March 1959.

\5 Matick, Richard E., "Transmission Line Pulse Transformers — Theory and Application," **Proceedings of the IEEE**, Vol. 56, No. 1, Jan. 1968.

\6 Hall, Gerald L., K1TD, ed., **The ARRL Antenna Book**, 14th ed., Chapter 5, (ARRL, 1982).

\7 In practice, there is always an RF path from the rig to ground, and its impedance should be made as low as possible. The rig should **always** be dc grounded for safety.

\8 See Ref. 6.

\9 Nagle, John J., K4KJ, "High-Performance Broadband Balun," **Ham Radio**, Feb. 1980, p. 28.

\10 See Ref. 1.

\11 Reisert, Joe, W1JR, "Simple and Efficient Broadband Balun," **Ham Radio**, Sept. 1978, p. 12.

Available Power, SWR and Loading

By David T. Geiser,* WA2ANU

*RR 2, Box 787, Snowden Hill Rd., New Hartford, NY 13413

The Power Transfer Circuit

Simple reasoning and easy arithmetic eliminate much of the mystery from SWR and loading. This approach can spark new thoughts in the novice and old-timer alike.

Any RF power-transfer circuit can be represented by a series connection made up of a generator, internal generator resistance, load resistance and reactance (Fig. 1). Remembering that the source may include an RF transmission line from a transmitter, the internal generator resistance may represent generator or source losses.

The power actually delivered through the output terminals goes to the load resistance. Notice that while power is delivered to the load, some of it is also dissipated to the source.

Any circuit that is not intentionally tuned has an overwhelming chance of showing inductive or capacitive reactance. This reactance (X) can be shown in series with the generator internal resistance (Ro) and the load resistance (RL).

Power actually delivered to RL is:

$$P_L = \frac{R_L E^2}{(R_L + Ro)^2 + x^2} \qquad \text{(Eq. 1)}$$

If we assume for the moment that we tune out the reactance (X = 0), delivered power PL is:

$$P_L = \frac{R_L E^2}{(R_L + Ro)^2} \qquad \text{(Eq. 2)}$$

We can work out an example to see what happens to the delivered power as we change the load, RL. Table I gives the values for E = 100 volts rms and Ro = 50 ohms. When we plot the results on semi-log graph paper (Fig. 2), we see a symmetrical curve with the power maximum at RL = 50 ohms. The power delivered is always a maximum when the generator and load resistances are equal and there is no net reactance. This amount of power is called available power. It would seem that this is an ideal condition; however, two other cases exist and are discussed in Appendix A.

Circuit Reactance

We must consider reactance because it is difficult to adjust X, shown in Fig. 1, to exactly zero.[1] Though a few ohms of reactance (compared to 50 ohms) do not cause much apparent difference, highly selective loads may display considerable reactance.

Let us plot resistance and reactance on a common graph, such that:

$$\frac{Ro}{R_1} = \frac{R_2}{Ro} = \frac{(R_2)}{(R_1)} \ (0.5) = SWR \qquad \text{(Eq. 3)}$$

Taking a tip from Table I, let us choose R1 and R2 with values of 25 and 100 ohms, respectively, and Ro equal to 50 ohms. Taking 25 and 100 as the end-points of a diameter (R2 - R1) of a circle (Appendix B), we can plot that circle (Fig. 3).

If we check several values of RL and X that lie on the circle's circumference, we find that the circle is a plot of all possible values of RL and X for a constant delivered power, calculated with Eq. 1. (Some check values to try for an SWR equal to 2 are: RL = Ro and X = 0.707Ro, or RL = 2Ro and X = 0). SWR is the relation that links the portion of the available power that can be delivered to an unmodified load to the relation between the actual and ideal loads.

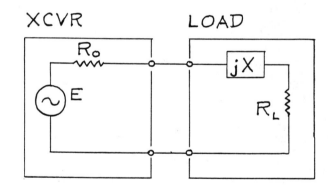

Fig. 1 -- Ideal RF generator with a partially reactive load.

Maximum Reactance

The impedance chart in Fig. 3 shows that for a specified SWR there is a maximum limit on capacitive and inductive reactance, 37.5 ohms for the SWR = 2 example. This inspires the thought that the reactance can be reduced to zero by series insertion of equal reactance of the opposite type. In the SWR = 2 example, reactance cancellation at X = 37.5 would reduce the SWR from 2 to 1.25.

If we find the SWR of a load with an SWR bridge, we can calculate the maximum possible reactance (Appendix B). The actual reactance type and magnitude will not be known, but a compensating circuit can be made (point A in Fig. 4). The values given are one of the many choices that would satisfy the SWR = 2 example.

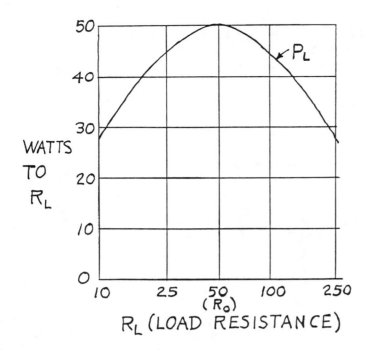

Fig. 2 — Power delivered from a 50-ohm source to various pure resistances.

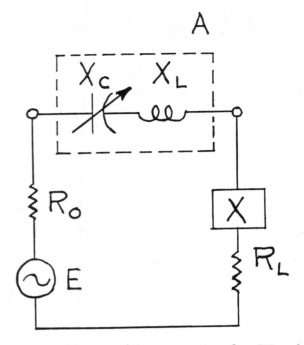

Fig. 4 — Simple series compensator for SWR = 2 in a 50-ohm system. There is no improvement for RL of 25 or 100 ohms.

Use of a series circuit near resonance makes the choice of a low-loss inductor (X_L) desirable. Even then, an increase in circuit resistance (from inductor losses) may result in less power being delivered to the actual load.

A mobile Amateur Radio operator can minimize his narrow-band antenna problems by loading the body of the antenna so it is slightly capacitive at the high-frequency band edge. A manual or motor-driven variable inductor at the feed point allows the antenna to resonate anywhere in the band.

While the mobile ham sees the resonated antenna as a low resistance, the base-station ham has many choices that allow the opportunity for minimizing the complexity of an antenna-matching circuit (Transmatch). When we look at the characteristics of a vertical antenna, we observe that the resistance at slightly more than a quarter wave (90 degrees) is close to 50 ohms, with different inductive reac-

tances (Table II).\2 Let's assume we chose a 95-degree length for 3.5 MHz — this gives us about 110 degrees at 4.05 MHz. Series capacitance compensation reduces the SWR to 1.3 or less across the whole band.

The experienced ham usually reminds me that you can also use parallel compensation for reactance, but the equivalent parallel resistance is not the same as series resistance (Appendix C). The lowest SWR would probably call for a length of 82 degrees at 3.5 MHz. In this case, the SWR is not as low as in the series example, but would be 2 or less between 3.5 and 4 MHz. The parallel choice also requires both inductive and capacitive reactance to cover the range.

Exact and Approximate Matching

The ham who wants to make an exact match will have to use a Transmatch (Appendix C), remembering that the L network is the most efficient. Unfortunately, when high ratios of RP/RS are encountered, the bandwidth of the L network can become very small. This is the reason the pi-L network is useful. I believe the broadest bandwidth is found when the Qs of each of its two L networks are equal, or, paraphrased, the ratio of Rp to the intermediate resistance is equal to the ratio of the intermediate resistance to RS.

An approximate match may be all that is necessary. Most transmitters behave well if the load SWR is 1.5 or less, seen at the transmitter antenna terminals. Three one-eighth-wavelength sections of coaxial cable and four T connectors are very useful tools.\3 Insertion of each of these sections between the load and the measuring point moves the measuring point on the impedance circle; four of them returns you to the starting point on the circle. Alone, they do not change the SWR significantly.

The procedure I follow is to cascade the sections (Fig. 5A). I start by connecting a variable

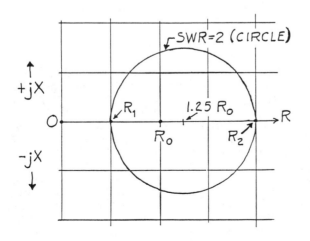

Fig. 3 — Impedance circle for all SWR = 2 in a 50-ohm system. (See text and Appendix B.)

capacitor to (1) and adjust it to lower the SWR. If the SWR changes, I leave it there and try a variable capacitor on (2), (3) and (4), respectively. If the capacitor at (1) did not reduce the SWR, I try it at the other locations. One position will reduce the SWR. Sometimes a second capacitor only increases the SWR. This means an inductance is necessary. A quarter-wave stub (Fig. 5B) will make a variable capacitor at (2) look like a variable inductor at (3), and losses are much less than in the conventional inductor. Once the SWR is reduced to an acceptable value (it is surprising how often it can be reduced to the ideal value of 1.0) any excess cable between the last used tap and the SWR bridge can be removed.

There is nothing unique about this form of compensation. Ideally, any load may be matched with two reactances spaced 1/8 wavelength on a transmission line. This method only minimizes precision measurements and cable-cutting.

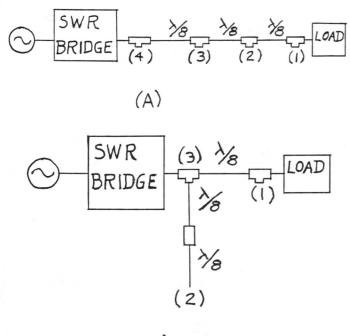

(A)

(B)

Fig. 5 — Using 1/8-wave sections for impedance transformation (A) shows a cascade arrangement of shunt-capacitor taps. At B is a transforming shunt capacitance to a low-loss shunt inductance. (See text.)

Appendix A — Intentionally Mismatched Loads

Efficiency and low distortion are two common reasons that a user may **not** want to feed a source into a matched load. Refreshing our memory of Fig. 1 and Table I, we note that the generator and load resistances were equal. The same current flowed through each, so each must, in the example, dissipate 50 watts. This is 50% efficiency.

Consider a generator of 60 V rms and an internal resistance of 10 ohms and zero reactance. A 50-ohm load still receives 50 watts, but the generator dissipates only 10 watts. This is an efficiency of 83%. Only 60 watts (instead of 100) has to be generated, and a smaller generator proves satisfactory.

Similarly, many generators deliver a better-quality (less distortion, harmonics, or noise) signal to a chosen mismatched load. The previous example choices (10-ohm source and a 50-ohm load) are not particularly unusual. Because third- and fifth-order distortion hurts SSB intelligibility and harmonics cause interference, a designed mismatch at the transmitter is often made. Similarly, the best noise-figure adjustment at the receiver input is often a power mismatch.

A mismatch at the load will usually appear different at the source if there is any distance between the two. Imagine the mismatch (in the 10-ohm example) changed to a "matched" 10 ohms. Now 90 watts are dissipated in the source and the source has to generate 180 watts instead of 60. Parts will often burn out. Yet this was a "matched" load. This sort of reaction (distortion, harmonics, noise, and damage) is dangerous when a high SWR is present.

We will call the 50 ohms the design load (not a matched load) and do our best to keep it at the resistive 50 ohms, or other design load value.

Appendix B — Rectangular Impedance Chart

The center of the circle, R, is located at:

$$R = \frac{Ro(1 + SWR^2)}{2SWR} \qquad (Eq. 4)$$

The radius, r, of the circle is:

$$r = \frac{Ro(SWR^2 - 1)}{2SWR} \qquad (Eq. 5)$$

Note that this is also the maximum possible reactance.

This is not the only kind of SWR circle that relates design loads to actual loads. The most common and useful is the Smith Chart, based on a logarithmic distribution of impedances. The Smith Chart places the design load at the center of the SWR circle (hinted at by the symmetry of Fig. 2) with phase angles proportional to angular rotation along the circle. It is more useful than the rectangular chart of this note, but harder to explain.

Appendix C — Series-Parallel Equivalence and Matching

At any single frequency, any two-terminal combinations of resistance and reactance may be expressed as either a series circuit or a parallel circuit. (Fig. 6.) We can make the circuit look purely resistive by series or parallel reactance of the opposite type. Then only the equivalent series or parallel resistance will be seen at that frequency (Fig. 7).

Here, there has been no change in the resistance. If, however, we use parallel compensation on the series circuit (Fig. 8) we see resistance RP instead of RS. Series compensation of the parallel circuit gives us RS instead of RP.

The reactances, XP, XS, -XP, and -XS, can be, but do not have to be, part of an antenna. They can be ideal inductors and capacitors.

Thus, with one capacitor and one inductor, we can transform any high resistance RP into a lower resistance RS. We first define a number, Q:

$$Q = \frac{R_P - R_S}{R_S} \ (0.5) = \frac{X_S}{R_S} = \frac{R_P}{X_P} \quad \text{(Eq. 6)}$$

From this simplification,

$$R_P = R_S(Q^2 + 1) \quad \text{(Eq. 7)}$$

$$X_P = X_S\left(1 + \frac{1}{Q^2}\right) \quad \text{(Eq. 8)}$$

$$X_S = QR_S \quad \text{(Eq. 9)}$$

$$X_P = \frac{R_P}{Q} \quad \text{(Eq. 10)}$$

Fig. 6 -- Series-parallel equivalence symbols.

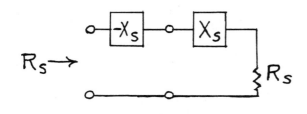

(A)

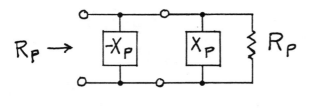

(B)

Fig. 7 -- Compensation. At A, series with series equivalence. At B, parallel with parallel equivalence. There is no change in equivalent resistance as seen looking into the circuit terminals.

If, for example, we wanted to transform 200 ohms (RP) to 50 ohms (RS),

$$Q = \frac{(200 - 50)\,(0.5)}{50} = \frac{(150)\,(0.5)}{50}$$

$$= (3)\,(0.5) = 1.732$$

$$X_S = 1.732(50) = 86.6 \text{ ohms,}$$

$$XP = \frac{200}{1.732} = 115.5 \text{ ohms,}$$

The resulting circuit is called an L network, as shown in Fig. 9. It is called this because of the usual relative positions of the reactances in a circuit diagram.

Pairs of L networks may be used for multiple impedance transformations; familiar types are the pi, T and pi-L. Relative impedances are indicated, and the usual harmonic-reducing configurations are shown in Fig. 10.

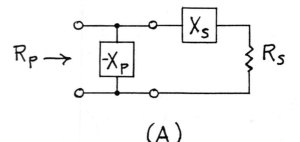

(A)

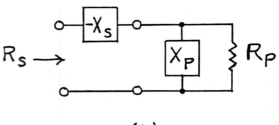

(B)

Fig. 8 -- Compensation. At A, parallel with series equivalence. RS transforms to RP. At B, series with parallel equivalence. RP transforms to RS.

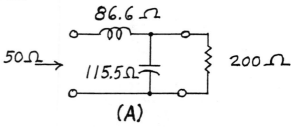

(A)

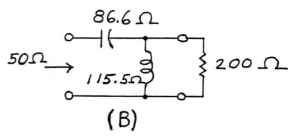

(B)

Fig. 9 -- Examples of L networks that transform 200 ohms to 50 ohms. At A, the series-inductance version attenuates higher frequencies more. The series-capacitance version at B attenuates lower frequencies more.

168

References

\1 Exact adjustment of reactance to zero (or another exact compensating value) may reduce distortion or harmonics a dozen or so dB. I have seen reductions up to 40 dB.

\2 The numbers in Table II apply to only one antenna, visually estimated from the graphs on page 2-24 of **The ARRL Antenna Book.**

\3 Remember that a wavelength of 50-ohm solid-polyethylene coaxial cable is only about 66% of a free-space wavelength. Other cables may have different velocity factors.

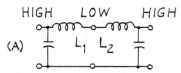

L_1 AND L_2 MAY BE COMBINED

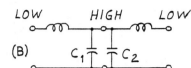

C_1 AND C_2 MAY BE COMBINED

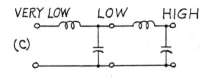

Fig. 10 — Transmatches. Pairs of L networks are combined to make a pi or π network (A), a T network (B), and a pi-L network (C). The "highs" at the two ends of the network at A are not required to be equal, and may differ by a ratio of 100:1 or more. The same applies to the "lows" at B.

Table I—Power Delivered vs. Load Resistance

LOW RL	HIGH RL	PL (WATTS)	SWR
50	50	50	1.00
40	62.5	49.4	1.25
33.3	75	48	1.50
25	100	44.4	2.00
16.67	150	37.5	3.00
10	250	27.8	5.00

Table II—Reactance Compensation of Vertical Antenna

Length (Degrees)	Series Resistance (RS)	Series Reactance (XS)	SWR	Series Compensated SWR
82	27	−35	2.96	1.85
85	30	−15	1.89	1.67
90	36	+17	1.67	1.39
95	40	+50	2.96	1.25
100	47	+80	4.51	1.06
105	56	+115	6.58	1.12
110	66	+150	8.78	1.32

Length (Degrees)	Parallel Resistance (RP)	Parallel Reactance (XP)		Parallel Compensated SWR
82	72.73	−55.8		1.45
85	37.5	−75		1.33
90	44.0	+93.2		1.14
95	102.5	+82		2.05

Mr. Smith's "Other" Chart and Broadband Rigs

By Roger K. Ghormley,* WØKK

*2115 South 24th St., Lincoln, NE 68502

Have you heard about the Smith Chart that is not **the** Smith Chart we know of today? This "other" chart actually came before the well-known one, and can play an important role in the ham shack. The following paragraphs show how this lesser known chart can assist today's amateur.

Mr. Smith

During the early 1940s, Phillip H. Smith was involved in the Radio Development Department of the Bell Telephone Laboratories, Inc. His chart (Fig. 1) was published as part of an article in the January 1944 issue of **Electronics.**\1 The other chart, actually a series of charts, was published in the March 1942 **Electronics** article, "L-Type Impedance Transforming Circuits." It covered eight variations, but we will look at only two of them.\2

With the increasing use of no-tune or broadband rigs, Amateur Radio operators are faced with the need for an "impedance transforming circuit." To illustrate the factors involved, we'll take a look at John Q. Ham. He is active in RACES (Radio Amateur Civil Emergency Service) and wants to be able to function on RACES frequencies in the 80- and 75-meter band — 3500 to 3550 kHz and 3984 to 4000 kHz.\3 Until now, there's been no problem. The tune and load controls of John's rig have been able to transform the impedance at the shack end of the transmission line to something that "satisfies" the final amplifier of his rig. But with broadband rigs, a problem develops.

The Problem

John Q.'s new modern broadband rig does not have a tune or load control. Why is this a problem? His rig was designed to "see" a resistance of 50 ohms and wants it fairly "pure" (not much reactance). If it does not see this, it will automatically cut back on output or blow the circuit breaker by going into oscillation! What's more, the shack end of John's transmission line (even though it is a 50-ohm RG-8 coaxial cable) shows an impedance that varies widely from 50 ohms, depending on the frequency. In fact, it is never a pure 50 ohms!

Fig. 1 includes a plot of how the impedance of John's dipole (125 ft 6 in of no. 14 wire) varies with frequency.\4 Two additional plots show the input impedance at the shack end of 40-ft and 75-ft lengths of RG-8 coax if terminated in John's dipole. (For now, ignore the "backward S" divider; it is discussed later.) All notes are included in the

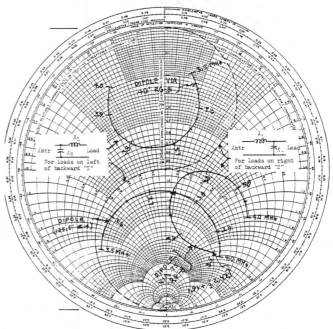

Fig. 1 — Plots of the input impedance of a dipole antenna (125.5 feet of no. 14 gauge wire) and of the input impedance for a 40-foot and a 75-foot RG-8 coaxial line terminated in the dipole over a frequency range of 3.5 to 4.0 MHz. In addition, the backward "S" shows the impedance ranges matched by the two types of L networks discussed in the text.

appendix.

Look at the "horseshoe" for John's 75-ft transmission line. The input impedance at 3.5 MHz is 1.91 + j2.42 ohms, normalized.\5 This is a pure resistance of 96 ohms in series with an inductive reactance of 121 ohms. That represents an SWR of 5.31:1 — a bit above the manufacturer's suggested limit of 2:1!\6

At a frequency of 3.75 MHz, the input impedance is 54 + j17. Not bad. The SWR reading is 1.4:1. That should satisfy the rig. But at 4.0 MHz, the SWR is back up to 5.2:1. It is clear that John has to do something if the new rig is to work near the band edges.

The L-Net Answer

Having eliminated the tune and load knobs, John wonders what is the simplest workable solution to the problem — without adding more panel knobs. Mr. Smith's "L-Type Impedance Transforming Circuit," the L network, now enters the scene.

The L net consists of two elements; a series and a shunt arm. This is about as simple as you can get! It transforms any complex impedance (resistance plus reactance) into a pure resistance of any desired value. As shown later, its tuning is quite broad, so it provides a workable match over a fairly good frequency range. Sound good? Let's see how John goes about making up such a net.

We can now look at Mr. Smith's "other" chart, which we will call the Smith L Chart (Figs. 2 and 3). Fig. 2 applies to an L net when the shunt arm is next to the load. I have divided this figure into A and B for clarity. Fig. 3 applies when the shunt arm is next to the transmitter. Which figure we should use is covered in a later section.

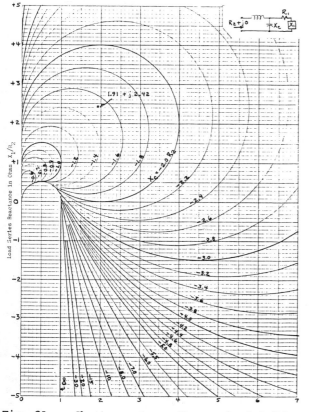

Fig. 2A — Shunt reactance X_C required (with Fig. 2B) to transform load impedance $R_1 \pm jX_1$ to a pure resistance $R_2 + j\emptyset$ ohms.

Using the Smith L Chart

Locate the point on Fig. 2 or 3 that corresponds to the normalized load impedance. Read the reactance (in normalized ohms) directly from the chart for the shunt and series arms of the L net. For example, a few paragraphs earlier we learned the input impedance to John's 75-ft line at 3.5 MHz was $1.91 + j2.42$ ohms. Find that spot on Fig. 2A (there is one there for your convenience), and read the required shunt arm reactance of 1.5 capacitive ohms. The same procedure on Fig. 2B gives a series arm reactance of 2.0 inductive ohms.\7

Convert these normalized values to -75 ohms for the shunt capacitance and 100 ohms for the series inductance. Multiply by 50 to satisfy John's rig. Now convert these ohms into component values for the capacitor and the inductor, either by using the formulas:

$$ C = \frac{1}{2\pi f X_C} \quad \text{and} \quad L = \frac{X_L}{2\pi f} $$

or your favorite reactance chart in **The ARRL Handbook** for 3.5 MHz.

The results at 3.5 MHz are summarized in Fig. 4. Note that the complex impedance $96 + j121$ ohms appears as a pure resistance of 50 ohms at the input to the L net. John's rig would like that!

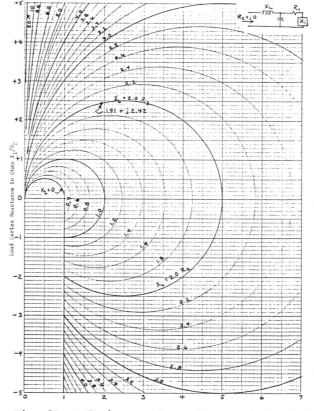

Fig. 2B — Series reactance X_L required (with X_C Fig. 2A) to transform load impedance $R_1 \pm jX_1$ to a pure resistance $R_2 + j\emptyset$ ohms.

The Backward S

An inspection of the Smith L Chart of Figs. 2 and 3 shows that Fig. 2 is used for all loads having a normalized resistive component greater than 1. For loads having a resistive component less than 1, the sign and size of the reactive component dictate whether Fig. 2 or 3 is used.

The same information in a different form is shown by the "backward S" of Fig. 1. The area in the right half is served by Fig. 2, and the area in the left half by Fig. 3. Had John's transmission line been 40 ft long instead of 75 ft, we would have used Fig. 3 to determine the L net. In fact, let's use Fig. 3 and assume the transmission line **is** 40 ft long. Try to find a way to satisfy John's rig over the entire 80- and 75-meter band with a minimum of knob twisting.

John's "Broadband" Dipole

Fig. 5 shows a plot of SWR vs. frequency at the shack end of 40 ft of RG-8 that terminates in John's dipole. The SWR at the band edges is now up to 5.7:1. This is because the 40-ft line has less attenuation than the 75-ft line (0.16 dB vs. 0.30 dB). Since John's rig works reasonably well with an SWR at about 2:1, he needs no matching network over the frequency range of 3.66 to 3.81 MHz.

Let's now see if we can find an L network to

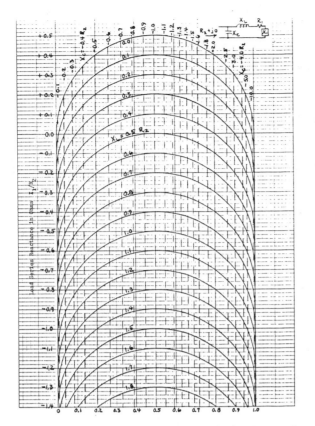

Fig. 3 — Series reactance X_L and shunt reactance X_C required to transform load impedance $R_1 \pm jX_1$ to a pure resistance $R_2 + j0$ ohms.

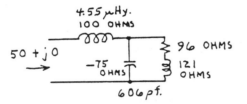

Fig. 4 — The L network, which transforms a load of 96 + 121 ohms into 50 + j0 ohms. Component values are for 3.5 MHz.

keep the SWR below 2:1 for the space below 3.66 MHz, and a second setting of the network for the space above 3.81 MHz. If so, then John can go from band edge to band edge with only two settings of the network, plus a "network out" position. Fig. 5 shows just that!

The L network is set for 50 ohms input at 3.57 MHz with an SWR of 1:1 in Fig. 5.\8 As the frequency goes down to 3.5 MHz, everything changes. The antenna resistive component goes down, the antenna reactive component gets bigger, the transmission-line effective length goes down, the transmission-line loss goes down, the L network shunt arm reactance gets bigger, and the series arm reactance gets smaller. As the frequency moves up to 3.66 MHz, all of these items change accordingly. Even so, the end result gives an SWR of 1.9:1 or less over the bottom portion of the band. This is shown in Fig. 5.

Similarly, an L network is fashioned for the top portion of the band. With the L network designed for a perfect match of 50 ohms at 3.905 MHz, the SWR is 1.82:1 at 3.8 MHz, and 1.95:1 at 4.0 MHz. With an

SWR of less than 2:1 across the entire 80- and 75-meter band, John's rig has a nice broadband system. He does it with an **ordinary** dipole and a bit of matching.

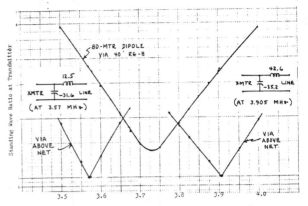

Fig. 5 — A plot of SWR vs. frequency for a dipole (125.5 feet of no. 14 wire) via 40 feet of RG-8 cable, showing taming of SWR by the use of two settings of an L network. Network values are in ohms for the frequency indicated. The input impedance of the line is shown as part of Fig. 1.

"However," You Say —

Are you thinking that John must have had a good RF impedance bridge to measure all those input impedances? How about those of us who have only an SWR meter of questionable accuracy? Good question. Here's what we'll do.

If we knew the largest values of shunt capacitance and series inductance we might have to use, we could "breadboard" a test network with confidence. We would then be prepared for whatever actual values are required during a trial-and-error procedure.

Fig. 6 shows the possible extremes in series and shunt arm values for an L network when only the SWR is known.\9 The extreme for the shunt arm is the **lowest** value of capacitive reactance; this requires a **large**-value capacitor. The extreme for the series arm is the **highest** value of inductive reactance; this requires a **large** coil.

These extremes are only possibilities, and they do not occur at the same time. They are required only if the transmission line length is such that the resistive component of the input impedance to the transmission line is either a minimum (requiring a large shunt capacitor) or a maximum (requiring a large series coil). A longer or shorter line reduces component values.

A Practical Case

Some years ago I acquired my first no-tune rig, a Triton IV. Appreciable operating was (and still is) done in the 80- and 75-meter band. (The same problem was also present for a tri-band beam, but the 80- and 75-meter case is the only one presented here.) The average of three dubious-quality SWR meters gave the SWR vs.frequency at the shack end of RG-8 transmission line as shown in Fig. 7. The simplicity of the L network was appealing, but no impedance bridge was at hand to help out. However, it looked probable that we would be liable for the largest component values for the L network down around 3.555 MHz, dealing with an SWR of about 4.5:1.

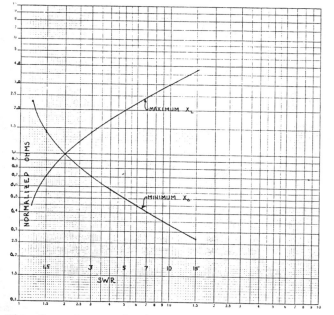

Fig. 6 -- Curves showing the minimum possible shunt arm reactance, X_C, and maximum possible series arm reactance, X_L, required for L networks terminated in lines of various SWRs.

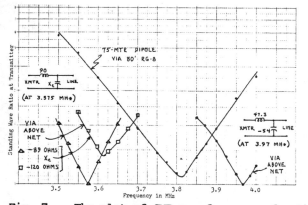

Fig. 7 -- The plot of SWR vs. frequency for the WØKK dipole (123 feet of no. 12 wire) via 80 feet of RG-8 cable, showing taming of the SWR by the use of three settings of an L network. Network values are in ohms for the frequency indicated. Input impedance to the line is unknown.

Entering Fig. 6 at an SWR of 4.5:1 yields a minimum liable X_C of 0.53 normalized ohm. That converts into a capacitance of 1690 pF at 3.55 MHz. Here was the first stumbling block. The largest transmitting-type capacitor at hand was 1000 pF. Perhaps luck was also at hand, and the maximum would be less than 1690.

From Fig. 6, the maximum liable X_L at an SWR of 4.5:1 is about 1.9 normalized ohm. That would be a maximum inductance of 4.3 uH at 3.55 MHz. No problem.

An L network was breadboarded as diagrammed in Fig. 8, using the capacitor at hand and some Minidux with taps. The network was placed between the SWR meter and the transmission line. With low power applied to the system, the network was adjusted by trial and error for a low SWR at all frequencies of interest for each antenna via the line. A record was

made of coil taps, capacitor settings, and whether the shunt arm was next to the load or transmitter. The 1000-pF capacitor proved to be large enough, though I had already decided to add transmission line if necessary to keep below this value for the shunt arm.

By using the ARRL Type A L/C/F Calculator [Discontinued — Ed.], the inductance values found for the breadboard coil were translated into a heavier-duty transmitting version, with some extra for error. The transmitting version of the completed network is a duplicate of the circuit in Fig. 8.

A perfectionist might want a roller-type inductor for the coil in order to get a "pure" match at all frequencies. The rest of us would be satisfied with the quick and easy convenience of about four coil taps plus "shorted" and "full-coil" positions. These fit nicely on the band switch taken from the network final of some defunct rig.

The end result for my 80- and 75-meter dipole plus L network is seen in Fig. 7. The lower portion of the band is not quite covered with a single setting of the network, and a twist of only the capacitor is required. That's good enough for me. Some day I might add a few feet to the dipole, to make it look almost as good as John's.

The Smith Chart is a powerful tool. Maybe all have described can help today's amateur benefit from Mr. Smith's "other" chart. It's one of the simplest, least expensive ways to get a broadband antenna system to match you broadband rig. Give it a try!

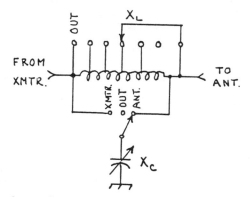

Fig. 8 -- The circuit of an L network is used for experimentally determining component size. The transmitting version is the same except for heavier-duty components.

Appendix

\1 Phillip H. Smith retired from the Bell Labs in 1970 and presently operates the Analog Instruments Company of New Providence, NJ. One of the best sources of information on the Smith Chart[R] and its use is his book, **Electronic Applications of the Smith Chart**, 1969, Robert E. Krieger Publishing House, Melbourne, FL. The book contains a number of overlays for use with a Smith Chart, including one that provides a single-chart method of determining the series and shunt arms of an L network for any impedance location on the chart.

\2 An L network may have either inductive or capacitive reactances in one or both arms. This results in eight possible configurations by moving the shunt arm next to the load or next to the power source. The series inductor/shunt capacitor styles used here were chosen because they include all load impedances and attenuate the higher harmonics between them.

\3 Palm, **FCC Rule Book**, 1983, ARRL Publication No. 47, p. 6-11, Table 1, RACES Frequencies.

\4 The impedances for Fig. 1 were obtained as follows:

a. Input impedance of the dipole itself was calculated from \10

$$Z_d = R + j[Y - 120 (\ln 24L/d - 1) \cot m] \quad \text{(Eq. 1)}$$

$$R = m^{2.736}/ 3048 \quad \text{(Eq. 2)}$$

$$Y = m^{2.234}/ 549.7 \quad \text{(Eq. 3)}$$

$$m = 180 (fl/984 + 0.013) \text{ degrees} \quad \text{(Eq. 4)}$$

L = antenna length in ft

d = antenna diameter in inches

f = frequency in MHz

b. The input impedance of the transmission line terminated in the impedance Z_d above was calculated from \11

$$Z_i \text{ (normalized)} = \frac{Zr \cosh vx + \sinh vx}{\cosh vx + Zr \sinh vx} \quad \text{(Eq. 5)}$$

$$Z_r = \frac{Z_d}{Z_o} = \frac{\text{Dipole impedance}}{\text{Trans. line } Z_o}$$

v = a + jB the propagation constant   (Eq. 6)

a = attenuation constant in nepers per unit length

B = phase constant in radians per unit length (2 π radians = 360 degrees)

x = line length in units

From **The ARRL Antenna Book**, 14th edition, p. 3-20, Fig. 34, the formula to determine RG-8 cable in dB per 100 ft

Loss at 1 MHz = 0.206 = a 1^x

Loss at 300 MHz = 3.9 = a 300^x

300^x = 18.932 (by dividing the second equation by the first) x = 0.5156.

From which,

Loss at f MHz = 0.206 $f^{0.5156}$ dB per 100 ft

Attenuation constant a = 0.02372 $f^{0.5156}$ nepers per 100 ft (1 neper = 8.686 dB).   (Eq. 7)

From **The Antenna Book**, p. 2-1

$$W = \frac{984}{f_{MHz}} \text{ ft} \quad \text{(Eq. 8)}$$

From **The Antenna Book**, p. 3-19, Table 1

$$W = 0.66 \frac{984}{f} \text{ ft for RG-8 coax} \quad \text{(Eq. 9)}$$

From Skilling \12 Eq. 1-48

$$W = \frac{2 \pi}{B}$$

Hence,

$$B = \frac{2 \pi f}{0.66 \times 984} = 0.9675 f \text{ radians per 100 ft at f MHz} \quad \text{(Eq. 10)}$$

\5 See Ref. 1.

\6 The SWR for Figs. 1 and 6 was calculated from \10

$$SWR = \frac{A + B}{A - B} \quad \text{(Eq. 11)}$$

where,

$$A = \sqrt{(R + Z_o)^2 + X^2}$$

$$B = \sqrt{(R - Z_o)^2 + X^2}$$

Z_o = Characteristic impedance of system

R = Resistive component of impedance

X = Reactive component of impedance

\7 The charts drawn in Fig. 2 include all impedance combinations represented by SWR values up to 7:1, and most of them to an SWR = 10:1. (Fig. 3 is readily extrapolated by inspection for values of X_1 off the chart.) For loads involving higher SWR values, the L network values may be calculated directly. The equations are developed as outlined in Figs. 9 and 10 and as follows:

a. In Fig. 9 (shunt arm next to the load as in Fig. 2), X_c is chosen to make $R_{EQ} = 1$. Then X_L is set equal to the absolute value of the resulting X_{EQ}.

$$X_c = \frac{X_1 - \sqrt{X_1^2 + (R_1 - 1)(R_1^2 + X_1^2)}}{R_1 - 1} \quad \text{(Eq. 12)}$$

$$\left|X_L\right| = X_c \frac{R_1^2 + X_1^2 + X_1 X_c}{R_1^2 + (X_1 + X_c)^2} \quad \text{(Eq. 13)}$$

X_c is a negative number (X_c = -n)

R_1 = load resistive component * ($R_1 \neq 1$)

X_1 = load reactive component (-n for capacitive and +n for inductive, where n is the numerical value)

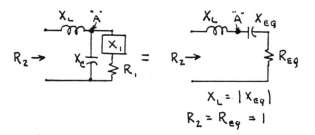

Fig. 9 — The development of the L network equations with the shunt arm next to the load. (See Appendix, note 7a).

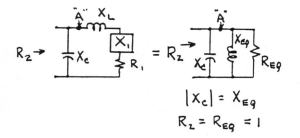

Fig. 10 — The development of the L network equations with a shunt arm next to the source. (See Appendix, note 7b).

R_1 and X_1 are normalized values

* If R_1 is less than 1, X_1 must be positive and equal to or greater than

$$\sqrt{R_1 (1 - R_1)}$$

For the special case where $R_1 = 1$.

$$X_L = X_1$$

For X_c

If X_1 is negative or zero, $X_c = \infty$

If X_1 is positive, use $R_1 = 1.00001$ in Eq. 12 to get around trying to divide by zero. Or, if you want X_c exactly, use Eq. 14.

$$X_c = \frac{-(3X_1^2 + 1) \pm \sqrt{(3X_1^2 + 1)^2 - 8X_1^2(X_1^2 + 1)}}{4X_1}$$

(Eq. 14)

In the numerator, use a negative radical for X_1 less than 1 and a positive radical for X_1 greater than 1. (The radical is zero at $X_1 = 1$). (This unlikely looking Eq. 14 is derived from Eq. 13 by setting $X_L = X_1$.)

b. In Fig. 10 (shunt arm next to the transmitter as in Fig. 3), X_L is chosen to make $R_{EQ} = 1$. Then X_c is set equal to the resulting X_{EQ}.

$$X_L = -X_1 + \sqrt{R_1 (1 - R_1)} \qquad \text{(Eq. 15)}$$

$$X_c = \frac{R1}{\sqrt{R1 (1 - R1)}} \qquad (X_c \text{ depends only on } R_1)$$

(Eq. 16)

X_c is a negative number ("X_c" = -n)

R_1 = load resistive component and is less than 1.

X_1 = load reactive component and must be either negative, or if positive, must be equal to or less than

$$\sqrt{R_1 (1 - R_1)}$$

c. Note that the semi-circle boundary between Figs. 2 and 3 is the locus of points where $X_L = 0$ and

$$X_1 = \sqrt{R_1 (1 - R_1)}.$$

Figs. 9 and 10 hence become identical.

\8 In Fig. 5, the frequencies at which to set the L network for a perfect match were found by trial and error to hold the SWR below 2:1 as the frequency was varied above and below the design frequency over the required range. The process makes heavy use of a programmable calculator!

\9 The reactance of the maximum-size **components** of Fig. 6 are given by the following:

$$X_{CMin} = \frac{1}{\sqrt{SWR - 1}} \quad \text{ohms} \qquad \text{(Eq. 17)}$$

$$X_{LMax} = \left| \frac{1}{X_{CMin}} \right| = \sqrt{SWR - 1} \quad \text{ohms} \qquad \text{(Eq. 18)}$$

\10 J. Hall, "The Search for a Simple, Broadband 80-Meter Dipole," **QST**, April, 1983.

Eq. 1 and its terms are taken from the Appendix of Hall's article. Eq. 11 is also taken from the same article as is Eq. 6.

\11 Skilling, **Electric Transmission Lines** (New York: McGraw-Hill Book Co.). Eq. 5-33, p. 88.

\12 Ibid., p. 15.